ENVIRONMENTAL PRESERVATION AND THE GREY CLIFFS CONFLICT

ENVIRONMENTAL PRESERVATION AND THE GREY CLIFFS CONFLICT

Negotiating Common Narratives, Values, and Ethos

KRISTIN D. PICKERING

UTAH STATE UNIVERSITY PRESS
Logan

© 2024 by University Press of Colorado

Published by Utah State University Press
An imprint of University Press of Colorado
1580 North Logan Street, Suite 660
PMB 39883
Denver, Colorado 80203-1942

All rights reserved
Printed in the United States of America

 The University Press of Colorado is a proud member of the Association of University Presses.

The University Press of Colorado is a cooperative publishing enterprise supported, in part, by Adams State University, Colorado State University, Fort Lewis College, Metropolitan State University of Denver, University of Alaska Fairbanks, University of Colorado, University of Denver, University of Northern Colorado, University of Wyoming, Utah State University, and Western Colorado University.

∞ This paper meets the requirements of the ANSI/NISO Z39.48-1992 (Permanence of Paper).

ISBN: 978-1-64642-574-7 (hardcover)
ISBN: 978-1-64642-575-4 (paperback)
ISBN: 978-1-64642-576-1 (ebook)
https://doi.org/10.7330/9781646425761

Library of Congress Cataloging-in-Publication Data

Names: Pickering, Kristin D., author.
Title: Environmental preservation and the Grey Cliffs conflict : negotiating common narratives, values, and ethos / Kristin D. Pickering.
Description: Logan : Utah State University Press, [2024] | Includes bibliographical references and index.
Identifiers: LCCN 2023033389 (print) | LCCN 2023033390 (ebook) | ISBN 9781646425747 (hardcover) | ISBN 9781646425754 (paperback) | ISBN 9781646425761 (ebook)
Subjects: LCSH: United States. Army. Corps of Engineers. | Conflict management—Tennessee—Case studies. | Communication—Social aspects—Tennessee—Case studies. | Environmental protection—Social aspects—Tennessee—Case studies. | Environmental management—Tennessee—Case studies.
Classification: LCC HM1126 .P525 2024 (print) | LCC HM1126 (ebook) | DDC 303.6/909768—dc23/eng/20230902
LC record available at https://lccn.loc.gov/2023033389
LC ebook record available at https://lccn.loc.gov/2023033390

Cover photo: iStock/Wachiraphorn

To Grey Cliffs and its Community: May the narratives and stories of positive change continue on.

CONTENTS

List of Illustrations   ix

Acknowledgments   xi

1. The Grey Cliffs Conflict: Situating the Issues   3
2. Narratives, Stories, Ethos Building, and Environmental Justice   24
3. Community Narratives and Ethos: Agency and Values   49
4. Motivating the Compliant Individual: A Corps Resource Manager's Rhetoric of Regulation   67
5. Attempting to Persuade as a Community Organizer: Norma's Narrative of Logic Without Emotion   83
6. A Corps Resource Manager's Rhetoric of Relationship: Co-Constructing Ethos With a Community   122
7. Narratives of Jointly Accomplished Social Action Through Aligned Values: The Negotiated Resolution   147
8. The Continued Negotiation Process: Implications for the Future   199

References   221
Index   231
About the Author   237

ILLUSTRATIONS

FIGURE

7.1.	Edwards's and the Community's Initial Values, Evolving Aligned Values	188

TABLES

1.1.	Annotated Research Timeline Totaling Approximately 18 Consecutive Months	17
3.1.	Coding Scheme for Community Values	51
4.1.	Rhetorical Framework for Analyzing Edwards's Narrative—Credibility	70
5.1.	Rhetorical Framework for Analyzing Norma's Narrative—Credibility	86
5.2.	Coding Scheme for Norma's Values	88
6.1.	Rhetorical Framework for Analyzing Edwards's Narrative—Character	126

ACKNOWLEDGMENTS

Many deserve sincere thanks and appreciation for helping this book become a reality. First, I thank David Edwards and the Grey Cliffs community for sharing their experiences, thoughts, and stories with me. They spent much time being honest with me about issues near and dear to all of their hearts. Their willingness to be vulnerable and to trust me, a relative outsider, with their treasured memories and hopes for the future resonates with me as a true honor. My hope is that I have represented all who invested so much into this research project with the accuracy and justice they deserve.

I also thank my family, specifically my husband, Scott; my daughter, Leslie; and my son, Johnny. They patiently tolerated my long years spent at the computer as I put my best efforts toward assimilating much research, data, narratives, and stories into a coherent piece of work. I thank Scott for all of the dinners he prepared and all of the dishes he washed. I thank all for not complaining while I dedicated so much time to this work. Without you all and your support, I could never have accomplished all that I have.

Colleagues and reviewers deserve thanks, as well. My department chair at Tennessee Technological University, Linda Null, is so effective and efficient at her job that she makes mine as a faculty member so much easier. I thank other colleagues, as well, for their friendship during this sometimes solitary, very focused time. The feedback the manuscript reviewers gave provided insightful and valuable guidance on shaping this work.

Finally, I express appreciation and thankfulness for Grey Cliffs. Although I live very close to it, I had no idea of its true significance and value until I got to know the case study participants, who generously shared their past memories and hopes for a positive future with me. The beauty of this area is astounding, its potential seemingly limitless. I am so glad I now understand a bit more about why so many spend countless hours appreciating this environmentally improved natural area. My hope is that this work will serve as a documentation of sorts, revealing how deeply so many care about it.

This book is an expansion of "Negotiating Ethos: An Army Corps of Engineers Resource Manager Persuades a Community to Protect a Recreational Lake Area," written by Kristin Pickering and published in *Business and Professional Communication Quarterly*. Copyright © 2021 by the Association for Business Communication. Reprinted by Permission of SAGE Publications.

ENVIRONMENTAL PRESERVATION AND THE GREY CLIFFS CONFLICT

1
THE GREY CLIFFS CONFLICT
Situating the Issues

David Edwards, a resource manager for the U.S. Army Corps of Engineers, had a problem. One of the county sheriffs in the district where he worked in the Southeast had given him 90 pages of documented crimes and disruptions that had taken place at a recreational lake area, Grey Cliffs, over the past 2 years. (All participant and location names within this study are pseudonyms.) Edwards immediately expressed concern because Grey Cliffs fell under his management responsibility. These nefarious activities included theft, drug use, kidnapping, attempted murder, assault, rape, and others. As Edwards monitored social media use about Grey Cliffs, he found warnings to people visiting the area, such as admonitions not to leave valuables in cars because thieves would break into vehicles and steal personal belongings. Upon visiting Grey Cliffs, Edwards found used needles, trash, and beer bottles littering the landscape surrounding the lake, which he attempted to clean up himself. In addition, he noticed evidence of all-terrain vehicle use that had decimated this once-beautiful area. Resulting erosion contributed to mudslides, and camping outside designated areas caused fires to burn dangerously close to trees and other vegetation. Trees riddled with bullets from target practice testified to continued unauthorized use of this land, as sportsmen prepared for upcoming hunting seasons. Grey Cliffs, one of 41 access points on this lake that the Corps managed, quickly had degenerated to become the very worst example of land management experienced in this area. Implications for this continuing environmental abuse and criminal activity were sobering; this area, ideally intended for public use, may need to be closed to prevent additional damage.

The Grey Cliffs community also had a problem. Recently, community members had heard rumors that the Corps might close Grey Cliffs, a beautiful area and beloved space that had served as the site for family swimming lessons, family reunions, fishing trips, picnics, cookouts, camping, hiking, baptisms, blackberry picking, Fourth-of-July celebrations, and, yes, all-terrain vehicle use. Families from this community had visited Grey Cliffs for generations and shared stories fondly of family time spent in this area, which was just down the street from where many lived. Considering this area their own, these stories contrasted sharply with others this community told about Corps land takeovers in the 1930s and 1940s, when the Corps created the lake to control flooding and generate hydroelectric power; many families had lost their farms that had been passed down from generation to generation during that time. To these community members, Grey Cliffs seemed almost like a consolation prize, an accessible area where they could be assured of convenient recreation opportunities. The community members weren't the only ones who valued Grey Cliffs. Kayakers, canoeing enthusiasts, and influencers praised the area on social media as a site for sporting activity and beautiful surroundings. Campers admired the lake view and rising grey cliffs that jutted upward from the lake, topped with lush trees and rock outcroppings. The area provided a sense of isolation so that, not too far from more populated areas, families could gather, recreate, and "get away from it all." Fishermen spent many hours on the lake catching catfish, walleye, black crappie, trout, and bass, journeying for miles if they wished or anchoring near the lake-access point in solitary coves too numerous to count. Because of community members' genuine, longtime love of this area, rumors of potential closure struck a strident, unharmonious chord with this community; they were angry even at the idea that the Corps would consider such action, and they weren't about to stand silently by and watch it happen.

Although not able to communicate with a human voice, Grey Cliffs as a physical location also revealed that it had a problem. While visited often by caring community members who did clean up the area after use, this area had also become known as a place where others could go to "get out of the eyes of the law." To everyone's best knowledge, no one monitored the area, and the area was so remote that even attempted surveillance seemed difficult, if not impossible. No one could even access cell service in the area. In addition, the muddy landscape visitors encountered simply could not continue as before, and the camping continued spreading far into the woods—far beyond what the Corps had intended. This site also became one that provided access to other people's properties that connected to the Corps land, providing opportunities for all-terrain vehicle users to trespass on others' properties. Grey Cliffs had witnessed much crime and environmental damage, and its future seemed sad and bleak. The activity of the area, some of it connected to family time and traditions and some of it crime related, all could very well result in restrictions that would prohibit anyone from accessing the area. These restrictions would certainly allow the area to rejuvenate in the quickest and most cost-efficient manner.

These various views and perspectives surrounding Grey Cliffs produced a kairos moment, a time when "the ability to select the right time and measure of language . . . a valuable rhetorical skill" (Salvo, 2006, p. 230) would impact this community and beloved, geographic space, perhaps forever. Edwards needed to take some action; the sheriff's reports were just one indication that activities at Grey Cliffs had gotten out of control. This community found itself in crisis and at a very difficult crossroads. Someone had to make some very difficult decisions, and no one was sure who would be making them. The community felt helpless as rumors spread, and the situation's urgency grew every day. Emotions escalated, anger spread, and conversations on front porches, back yards, and street corners grew more pointed within the community. [No one wanted to see this area closed, especially with its close community connections.] Motivation for action was quickly generating strength among community members as these conversations continued, but where would this building energy lead? Many of these community members harbored suspicions about anyone connected to the Corps and any community members who might be Corps sympathizers, who might be willing to restrict Grey Cliffs' access in support of the Corps.

[The environmental degradation and criminal activity, though, were clearly unacceptable, according to the Corps. Grey Cliffs had obviously become unmanageable; of all of the lake-access points, Grey Cliffs was by far the most notorious and crime ridden. Not only had the area sustained environmental damage, but human safety continued to be a growing concern. Even some of the local people voiced concern about visiting the area alone or at night.] Something had to be done to remedy these actions, and the Corps seemed to be the entity to step up and take control; after all, it did own the land and was in charge of maintaining it. Rumors continued circulating about the Corps closing the area. In order to begin a conversation about these issues, Edwards began talking with some local community members, who suggested a public meeting, one of several, to discuss the implications of closure. One of these community members was Norma; she lived near Grey Cliffs and was motivated to organize the first town hall meeting. She had experience with grassroots organizing and wanted to volunteer that experience to help the community.

At the first town hall meeting, Edwards and the community presented polarized narratives and views on Grey Cliffs' future status that ultimately reflected differing values. Edwards based his values upon the

Corps mission and vision, as he stressed the crime and environmental problems that no longer coincided with Corps goals. The community drew its values from Grey Cliffs' experiences as well as other values rooted in community traditions. The resulting narratives these opposing parties promoted were decidedly different as well; Edwards's narrative contained statistics from the sheriff's reports he received as well as his own experience with the area. The community's narrative contained testimonies about their use of the area, as well as clear resistance to closure. The community's narrative also focused on the benefits of the area and the inconvenience of closing it: no other lake-access points existed for miles. Based on the first meeting, Edwards and the community could not have been more polarized in their communication about Grey Cliffs. Nothing substantial could be accomplished without some type of value alignment and negotiation among these polarized communicators' narratives; something had to be changed to aid in resolving this crisis situation. How could change be accomplished, though? It is at this point that this observational case study begins.

RESEARCH QUESTIONS

This book presents an ethnographic, observational case study, including interviews that I conducted on communication and the events surrounding the Grey Cliffs lake-access conflict negotiation process. This study applies the theoretical lens of rhetoric (specifically the co-construction of ethos) to explore and articulate the relationship between ethos building and narratives, values, and texts, particularly when resolving conflicts. Specifically, this study addresses the following research questions:

- How do key participants' narratives reveal negotiated ethos, values, and action during the Grey Cliffs events?
- How do different participant values motivate attempts to negotiate action during this process, especially surrounding sustainability?
- How is ethos co-constructed among participants and articulated through texts?
- What persuasive strategies during this conflict appear to be failures and why?

Focusing on these research questions profiles the co-construction of ethos, values, and negotiation efforts illustrated in this case study, and the findings reveal ways narratives, conveyed through various texts, enable and/or constrain agency and ethos negotiation. This negotiation is essential for effective relationship building.

PURPOSE OF THE STUDY

My purpose in writing this book is to explore the research questions in light of the overarching concept that organizations and communities cannot negotiate meaningful action and relationships until there is a shared narrative that reflects aligned values. Constructing that shared narrative is the complicated part, and no one process is the same or works for all parties involved. As Faber (2002) writes when discussing a conflict he observed,

> Here, change was all about stories, but because the stories were so divergent, so opposite to each other, there was no possibility that either side was about to change. Instead, those in each group simply reinforced the other group's stories and perceptions held of their opponents. No one had created or presented a larger story to pull these people together; there was no common narrative they could both embrace. As a consequence, without a unifying story, one that spoke to both groups, neither side was about to change. (p. 8)

The Grey Cliffs conflict was similar to the one Faber discussed because both parties, the Corps and the community, promoted divergent stories and narratives; the Corps represented narratives of authority, poorly kept regulations, and crime, and the community communicated narratives of family gatherings, camping, and recreation. These types of conflict are common among organizations, businesses, and communities, and more research is needed to develop ways to create unifying stories and narratives among diverse groups. As Smith et al. (2020) propose,

> Future research should look at how organizational discourses around organizational innovation and failure may shape over time and the role communication plays in altering associated frames. Furthermore, it would be useful to understand the ongoing consequence of communication for shaping the understandings of what innovation work actually entails. (p. 20)

Far from being an idealized account of compromise for needed, innovative changes to Grey Cliffs' activities, however, this book presents an analysis of this particular organizational and community conflict and ways common narratives began to develop organically and realistically, based on these communicators' unique characteristics, motivations, and values. These common, negotiated narratives and values are just the beginning of conflict resolution and most likely will change over time, but studying the beginnings of this conflict resolution, as I do here, suggests tangible examples, attitudes, strategies, and frameworks for conflict negotiation that readers could apply in a variety of situations, whether as organizational members, community members, or participants functioning to various degrees within these realms. "Engaged in a shared activity"

(Smith et al., 2020, p. 2), these community workers demonstrate how "innovation is constituted through everyday talk and interaction" (Smith et al., 2020, p. 2). As such, this research helps answer the call for more inclusive research, particularly in the field of technical and professional communication, that has often suffered from a "hyperpragmatist" view (Scott et al., 2006, pp. 7–17) in the past. Instead, this more inclusive view "intentionally seeks marginalized perspectives, privileges these perspectives, and promotes them through action" (Jones et al., 2016, p. 214). Another important goal of this work is illustrating ways everyday practices, surrounding the shared activity of preserving access to Grey Cliffs, transpired through addressing the conflict. Eventually framing this event not just as a conflict but as opportunities for possible future action allowed Edwards, the Corps, and the community to work together on co-constructing solutions to these social and environmental problems.

BENEFITS AND UNIQUENESS OF WORK

This research extends work in the fields of ethos development, sustainability, values alignment, and narrative as it

- addresses the different sensemaking frames between a government organization and a rural community—the Corps needed to bring the community back into alignment with sustainability values, and the community needed to co-construct and revise its ethos with the Corps in order to negotiate access to Grey Cliffs;
- emphasizes co-constructed framing processes as a way to align values and actions discursively through all participants' narratives;
- connects organizational, environmental, and rhetorical communication theory and practice with cultural narratives, an application that potentially addresses other types of unique, organizational conflicts;
- highlights the relationships among ethos, value alignment, and shared identity development through co-constructed framing and rhetorical strategies, meeting a growing cultural need for additional research into accomplishing social action among participants with polarized views;
- illustrates how values and rhetoric can be adapted to the needs of a local culture with the aim of accomplishing common social action, extending the research on "our responsibilities to the cultures and communities within which, to whom, and about whom we communicate" (Haas & Eble, 2018, p. 12);
- provides data that support an increased understanding of why and how audiences change their actions based on persuasive discourse and socially mediated action regarding environmental and safety issues, based on negotiated ethos development;

- demonstrates the *dialogic* (Bakhtin, 1983; Meisenbach & Feldner, 2011, p. 567; Olman & DeVasto, 2020, p. 17) and *poly-vocal* (Boje, 2008; Jones et al., 2016, p. 212) work of organizational and community rhetors who, through rhetorical persuasion as well as agency and identity negotiation, work together to accomplish Corps environmental sustainability goals;
- promotes inclusivity by "working in communities and the public sphere" (Jones et al., 2016, p. 217) to learn more strategies for engaging with dominant narratives, such as organizational ones; and
- presents analysis of a generous range of texts used to accomplish and mediate communication goals through the qualitative, ethnographic, observational case study approach.

Many studies in organizational communication have focused on errant companies that damage the environment for capitalistic purposes or function shortsightedly, with seeming disregard for the concerns of impacted communities (Boyd & Waymer, 2011; Henderson et al., 2015; Jaworska, 2018; Lehtimäki et al., 2011; Shim & Kim, 2021; Verboven, 2011; Waller & Conaway, 2011). Many of these businesses address global markets and also serve a wide customer/company base. But this study is different in that this time the "big business"—the government—is the one prompted to step up to address a damaging environmental situation, and the community is the one to resist. Because of different cultural "sensemaking" (Weick, 1995) frames—including new materialist, embodied ways a person experiences physical and cultural events (Frost, 2018, p. 25; Herndl et al., 2018, p. 87; Senda-Cook et al., 2018, p. 102)—the Corps and the community approached this communication process differently. The cultural-historical context provides a unique setting through which to view the social negotiation of action between the community and the government: in the 1930s and 1940s, the Corps had bought much land surrounding a local river, including many farms, to create the lake to manage flooding and generate hydroelectric power. While this management undoubtedly had its benefits, many landowners at the time believed they had not been given a fair price for their land, they were not given a choice in whether to sell, and their generations-owned properties were lost. As a result, many landowners and farmers left the area to start over elsewhere (Williams et al., 2016). This cultural history led to a long and unpleasant narrative between this local community and the Corps, a narrative that had little if no positive history to balance it out. Community narratives had framed this history as an "us versus them" standoff that predisposed many community members to distrust government representatives in general and Corps representatives, such as Edwards as a resource manager, specifically. To add to this difficulty, there was no personal, embodied "face" to the Corps. Instead, it was an anonymous, unidentifiable entity that had suddenly reemerged to once again take control of this physical place the community felt it had regained some ownership of through memories and longtime, embodied recreational use. This scenario, then, is a very unique, local context through which to view the concepts of narratives, ethos, and values, all in need of some type of alignment in order for positive, social action and relationships to take place.

Organizational leaders in particular often use rhetoric to influence their audiences (Cheng, 2012; Heracleous & Klaering, 2014; Higgins & Walker, 2012; McCormack, 2014), and scholars have indicated the need to focus specifically on "ethos as an aspect of context that can shape rhetorical strategies" (Heracleous & Klaering, 2014, p. 133). The concept of ethos is a complicated one, including aspects of character development as well as expertise and authority (Aristotle, ca. 367–347, 335–323 B.C.E./2019). Only through ethos development can organizational leaders, for example, begin to negotiate action with the public, such as community members and stakeholders, who need to develop confidence and trust in the leader. In addition, community members need to develop their own ethos that complements the leader's ethos; only through this co-constructed ethos process can significant social action take place to resolve a conflict such as this one, which involves diverse members of the public.

In addition, this study contributes to the growing research focusing on environmental sustainability and organizational communication, including specific focuses on values and rhetoric adapted to the needs of a local culture with the aim of accomplishing common social action. It addresses needs of individual, rural, community stakeholders who are incredibly valuable to and legitimate in negotiating social action with an organization such as the Corps, although these community members might appear to be less significant and powerful at first, when compared to organizational communicators who have more ready access to dominant discourses of power. Edwards's ethos appeals, particularly those highlighting his credibility and character, reveal ways ethos can diplomatically frame an organizationally strategic message. Today, we are experiencing more and more tension between government organizations and the public. Analyzing the ethos creation of a government representative in a crisis, as well as this community's negotiated response to it, yields data and observations that scholars and communicators can use in thoughtfully and intentionally negotiating social action within different sociocultural contexts, including communities and governments, both influenced by discourses of organizational power (Bourdieu, 1986, 1990; Foucault, 1980, 1983, 1995). In addition to recognizing powerful organizational structures, this study also emphasizes the potential for community empowerment through "activism, social action, and the demarginalization of nondominant groups" (Walton et al., 2019, p. 109).

Recent work suggests that individual and community voices can indeed be heard and need to play a more meaningful part in corporate social responsibility, strategic communication, and organizational issues management (Carlson & Caretta, 2021; Henderson et al., 2015; Shim & Kim, 2021) as well as environmental communication and public policy formation (Carlson & Caretta, 2021; George & Manzo, 2022; Herndl et al., 2018; Le Rouge, 2022; Martinez, 2022). My work adds to this conversation by extending and further emphasizing the dialogic (Bakhtin, 1983; Meisenbach & Feldner, 2011, p. 567) work of organizational and community rhetors who, through rhetorical persuasion, attempt to work together to accomplish Corps environmental sustainability goals. This work also exemplifies an "ideal/real tension" (Meisenbach & Feldner, 2011, p. 566) that highlights potential strategies for negotiating communicative agency between organizations and individual stakeholders. The "ideal" Corps' regulation of the area differed significantly from the real lived experiences of the community members. The qualitative analysis of community participant interviews provides insight into the deconstruction of this dichotomy through the agency and identity negotiation process among the community, the Corps representative, and ultimately the Corps itself. Highlighting these community voices demonstrates a commitment to democratic communication about the environment; as Killingsworth and Palmer (2012, p. 265) discussed in their research, organizations must "recogniz[e] the need of all levels of people to have access to reliable information designed to be useful for their particular social goals" despite those goals being as seemingly insignificant as primitive camping and blackberry picking. Such attention supports "valu[ing] knowledge as experiential and lived" (Walton et al., 2019, p. 107), an important part of valuing the participation of marginalized communities as well.

Furthermore, Edwards's reflective observations on his own communicative processes contribute insights into these complex communication choices often not available from organization representatives in retrospect; this rhetor reveals that the identification process is nonlinear and recursive (Pickering, 2018) and is "key to how we perceive the world, looking through the lens of **historicity**" that occurs within a context (Jones & Walton, 2018, p. 242, bold emphasis in original), in this case, the context of the Grey Cliffs conflict, including its local community and narratives. When discussing social and environmental concerns, Higgins and Walker (2012) stress that discourse analysis alone can sometimes overlook "how other social actors think, feel and act" (p. 196) when discussing "social and environmental reporting" (Higgins & Walker, 2012, pp. 195–196). Discourse analysis alone therefore leaves a huge gap of missing information that analysis of reflective self-narratives can fill regarding social actors' "think[ing], feel[ing] and act[ing]" (Higgins & Walker, 2012, p. 196). My work provides data that contribute to an increased understanding of specifically why and how audiences, such as the Grey Cliffs community, change their actions based on persuasive ethos development initiated by an organizational communicator, "includ[ing] discussions of the practical implications of technical information, consistent efforts to make information accessible to the public, and a forthright representation of scientific uncertainties associated with complex technical information" (Tillery, 2006, p. 325).

The conflict analyzed here illustrates this local, cultural context as well as ways these participants used various rhetorical resources to negotiate agency and act. The community—those with less power in this story—did not have automatic and totalizing power bestowed on them similar to that seemingly possessed by Edwards as a Corps resource manager, a government representative. As a Corps representative, Edwards was charged with enforcing the Corps' identity as an organization. "An organization's identity or image is the result of an effort to create hegemony—the appearance of uniformity and consensus" (Graham & Lindeman, 2005, p. 423). Yet, once the community learned of the Corps' intentions to close the area, it began subverting and destabilizing those power structures and sense of order through the use of *polyphony* (Bakhtin, 1984), *heteroglossic* narratives (Bakhtin, 1983), *counterstory* (Martinez, 2020), and *antenarrative* (Boje, 2015); Edwards, in response, sympathized with the community's needs and negotiated with the community, in part to help neutralize antigovernment sentiment that had been generated between the community and the Corps, originating from the earlier Corps land buyouts. This process, then, reflected a negotiation among all parties; they learned the structures of the others' modus operandi and then acted with rhetorical awareness to subvert those structures and accomplish social action and agency within them (Giddens, 1984).

MY ROLE

My interest in this topic is personal as well as professional. I moved to a small farm that is part of this community about 15 years ago. Not being from the area, I have considered myself an "outsider" among people who have grown up in this rural community, where generations have raised their families. At the same time, I quickly realized I was also an insider since our farm is one of two that possesses direct access via rough, unpaved roads to the Corps land that borders the Grey Cliffs lake-access. In fact, to access a part of our farm, we must cross a small segment of Corps land. Via this rough road, "trespassers" from the Grey Cliffs lake-access area often ride all-terrain vehicles (ATVs) through the Corps land up to our house, which is at the top of the road leading down to the Corps property. I put "trespassers" in quotation marks because we have not treated them as such; they are not bad people or criminals but simply those looking for adventure who have wandered a bit too far. Nevertheless, they should not even ride ATVs on Corps property to begin with, which they would have to do before arriving at the bottom of the road that leads to our home. The presence/appearance of these people at our back door, despite "no trespassing" signs on our part of the road, emphasized this problem of unauthorized use of Corps land to us in a very personal way.

While we were uncomfortable with strangers appearing feet from our door at any time of day or night, we were not automatically in favor of closing the Grey Cliffs area to limit people from accessing it. One reason why was that we also used the area to take our children swimming and launch a small boat to fish and explore. We enjoyed using the area for what the Corps had intended; we also empathized with this community and wanted to support it in every way we could. We had heard of the community's devotion to the area and the history with the Corps, since we lived so close to the lake; we had met and gotten to know some neighbors, and our children went to school with others who were descendants of the original landowners in the area. As a result, we found ourselves occupying an in-between space in this conflict; we wanted to secure our own land (through closure), but we also wanted to preserve the use of the area for ourselves and the community. In this way, I realized my own unique positionality in relation to this growing conflict; my roles as community observer and participant were somewhat powerful as a landowner yet also weak since, ultimately, our family might not have any say in the Corps' decision about whether to close the area. Within this context, I realized the many selves that not only I but others participating in this conflict were constantly navigating and negotiating, through "differing ways of talking and being that stand as '[we]' for different audiences" (Gergen, 2007, p. 120). As Bourdieu (2007) emphasizes, "Because any language that can command attention is an 'authorized language', invested with the authority of a group, the things it designates are not simply expressed but also authorized and legitimated" (p. 170). As I observed this conflict, I saw how two different sides—the Corps and the community—were focused on legitimizing their own "authorized language" made up of scientific, governmental language on one side and affective, valued experiences on the other, and these dynamics created a rhetorical situation that everyone could learn from, including myself as a researcher.

When this situation unfolded in our backyard, I realized from an academic/professional perspective that this was a stunning example of various forms of communication: personal narratives, values, government/organization communication, business communication, environmental sustainability, other forms of rhetoric, and uses of various forms of texts. Given that these have been my scholarly focuses during my life in academia, I decided to become involved in this situation as a participant/observer, more an observer than a participant, since I still considered myself an outsider and was somewhat neutral about whether the area should be closed or not. In this vein, I began an ethnographic, observational case study involving various texts, including interviews, field notes, documents, and community stories that emerged from the interviews. With these data and observations, I hoped to explore the research questions through the context of co-constructing ethos, values, narrative, and texts, which I have found very useful in illuminating communication nuances involving conflict within these different cultural contexts. In doing so, I acknowledge that "the research process is itself a storytelling process in which the researcher's voice is always present" (Jørgensen, 2015, p. 285) throughout the presentation of others' voices, the theories I apply, and the data I've chosen to include. This information is crucial for learning more about conflict communication and ethos negotiation during crisis situations, especially involving sensitive cultural contexts, diverse populations, and the narratives they produce. As other scholars have demonstrated (Higgins & Walker, 2012; Mackiewicz, 2010; Walton, 2013), this type of specific analysis of rhetorical strategies provides an in-depth perspective on statements rhetors use when attempting to negotiate a credible ethos with audiences. Because I addressed the process of negotiating a shared identity with the audience using values as well, a grounded theory and interpretive approach allowed me to discover that Edwards's and the community's specific uses of language, in various forms, were part of the process of negotiating ethos among all participants.

METHODOLOGY AND DATA
Methodology
My research is based in qualitative (Denzin & Lincoln, 2018; Guba & Lincoln, 1994), ethnographic research techniques (Dunn, 2019; MacNealy, 1999; Strauss, 1987), including fieldwork practices, which highlight "identifying and assigning meaning by identifying participating actors, enabling attention to the mundane, and interpreting relevance with regard to rhetorical purposes and outcomes" (Grabill et al., 2018, p. 195). Grasping these community members' and resource manager's stories and "lived experiences" (Boussebaa & Brown, 2017, p. 14; Moore et al., 2021) was of the utmost importance in studying these communicators' rhetorical purposes. As Gephart (2007) asserts,

> Narratives and stories form the substance of much regulated communication; hence, narrative/rhetorical analysis addresses the *substantive* dimension of regulated communication and the form it takes. Rhetorical analysis complements narrative analysis by showing how selective construction of storytelling influences or regulates understanding and meaning. (p. 240)

This study highlights stories told by community members; these stories constitute reflective self-narratives about the conflict, its status, and the community's relationship to it. These narratives "can be a valuable method for sharing the individual and situated concerns of community members" (Stephens & Richards, 2020, p. 8), which may not be identified otherwise. As a result of the crucial importance of reflective self-narratives to learning more about these dynamic communication processes and negotiated power/action within this conflict, I incorporated semistructured interviews (Kvale, 1996) and allowed the participants to guide the discussion although I did have questions planned, the same for all interviews, to begin the discussion. My presence at the town hall community meetings allowed me to be a participant by "being there" (Blair, 2001; Rai & Druschke, 2018, p. 4; Senda-Cook et al., 2018, p. 103) and gathering additional background information to contribute to my general sense of the cultural history surrounding these events and conflict. At the same time, while being there, I understood that I can never fully and truly portray an authentic representation of participants' feelings, experiences, and motivations based on these observations because I did not live those same experiences. Likewise, this ethnographic portrait is a small snapshot of the events going on at the time (Heath, 1983); events and circumstances are constantly changing, just as the *sensemaking* (Grant, 2015, p. 113; Hargrave & Van de Ven, 2017; Henderson et al., 2015, p. 14; Weick, 1995) and embodied understanding (Frost, 2018, p. 25; Herndl et al., 2018, p. 87) of these events will not always remain stabilized. New dynamics can cause different conflicts to erupt, and negotiation may need to reoccur and evolve. The relationship between the community and the Corps will continue to change.

When analyzing the interview transcriptions, I implemented a grounded theory (Corbin & Strauss, 2015; Glaser & Strauss, 1967; Kwortnik & Ross, 2007) and interpretive approach (Heracleous & Barrett, 2001; Kuhn, 2006; Kvale, 1996) to identify themes significant to the theoretical lens of ethos building. As I read the interview transcripts several times, I made note of possibly significant recurring themes and then made connections among themes revealed by all of the interview participants. In the process, through open and in vivo coding (Saldaña, 2016), I selected key words, phrases, and sentences that reflected the application of those key themes; some of those quotations are included at chapter beginnings and through headings, for example, to illustrate participant agency and identity. Therefore, I viewed participants' words and collected data as an opportunity to discover efforts to portray and negotiate identities, based on the rhetorical exigency of kairos moments these participants deemed significant as they participated as social actors through these narratives. The resulting study yields a rich account of the messy process of negotiating action and relationships based on two general groups—the Corps and the local community—both with important, yet different, investments in the Grey Cliffs area.

Data

This ethnographic, observational case study includes qualitative data such as field notes taken during town hall meetings, documents distributed at meetings, maps, and semistructured interviews with key participants involved in the discussion and resolution of the issue. As Table 1.1 indicates, some data necessarily duplicated themselves in the field notes as well as the transcribed interviews. For example, Edwards stated his position title as Corps resource manager as he began his presentation at the first town hall meeting, and he reiterated his position title during our interview. Similarly, he reflected on statements he made at the meeting during the interview. The rich data I gained from analyzing the self-narratives are especially helpful in revealing participants' reflections on the events I observed at the meetings as well as other conversations that occurred before and after the meetings. The transcribed interviews revealed perspectives, intentions, and contexts that simply were not available from my observations alone. I transcribed all interviews by hand with the help of Express Scribe to ensure the transcriptions were as accurate as possible.

Table 1.1. Annotated Research Timeline Totaling Approximately 18 Consecutive Months

Research phase	Event	Date of event	Types of data collected
Pre-research	Informal conversation occurred between Edwards and a few community members at Grey Cliffs about the problems occurring there. All parties decided a community meeting would allow opportunities for discussion about these problems.	September 2018	None—researcher learned about these conversations during later community meetings and interviews.
Data collection	First community town hall meeting	October 2, 2018	Field notes, meeting agenda, meeting notes supplied by meeting organizer, Convention of States Literature
Data collection	Second community town hall meeting	October 16, 2018	Field notes, meeting agenda, meeting notes supplied by meeting organizer
Data collection	Meeting during which community members signed up for various committees	October 27, 2018	Field notes, meeting agenda, committee lists
Data collection	Joint committee meeting to update the community about progress being made on committee efforts and future plans	November 5, 2018	Field notes, negotiated action plan containing Corps and community goals, list of what help the Corps was willing to provide, email documentation from Edwards about plan for rejuvenation
Data collection	Joint committee meeting to discuss continued progress with helping to clean up the area through cleanup days, barricade installation, concrete and rock application	November 27, 2018	Field notes, revised map developed by Edwards indicating plan of future action for restricting access to certain areas, list of supplies to be provided by the Corps
Data collection	Participant interviews with Edwards, Norma, and selected community members	January 2019– March 2019	Recording of interviews
Interview recording transcription	Transcription by hand with the help of Express Scribe	Summer 2019	Interview transcripts
Data analysis	Analysis of all data collected during meetings and participant interviews	Fall 2019– Spring 2020	All data collected

© 2021 by the Association for Business Communication. Reprinted by Permission of SAGE Publications.

I initially recruited interview participants based on the prominence of their roles during the conflict. For example, because Edwards was the Corps resource manager in charge of informing the public about Grey

Cliffs' status and guiding future changes, I chose to interview him; he obviously knew a lot about this issue from his perspective and held an important role in addressing this conflict. I interviewed Norma because of her efforts to organize the community. Community participants also played strong roles, and I recruited some of them, such as Tom, who chose to lead some of the meetings and created a Facebook page for Grey Cliffs, and Paul, the owner of the general store where the town hall meetings occurred. A couple of participants, such as Denise and Felicia, serindipitously participated because they happened to be around when I was interviewing another participant and agreed to participate as well. All participants signed informed consent agreements for me to use their interview transcriptions in this research, and the overall research project was approved by the Institiutional Review Board at my university (Approval Number 2096). All identifying information has been removed from the data.

Some of the research I discuss in this study refers specifically to organizational communication within workplaces. While Edwards is an organizational representative, the community members are not workplace communicators within an organization. This characteristic contributes to the uniqueness of this study: not all organizational communication takes place within formal workplaces, and community members' reciprocal communication with organizations also proves essential when organizational efforts intersect community cultures and activities. The research I cite in this study yields perspectives that apply to these external stakeholders, community members, and workplace communicators. While not organization members themselves, these community members were still influenced by organizational communication through the Corps and Edwards in potentially life-changing ways. They also participated in co-constructing this communication, therefore participating in the power dynamics discursively constructed through that communication.

CHAPTER PREVIEWS

Chapter 2, "Narratives, Stories, Ethos Building, and Environmental Justice," presents the theoretical framework guiding the case study, including a focus on narratives and stories, as well as ethos and environmental justice. Within the context of narratives and stories, I introduce the concepts of symbolic and social capital, as well as agency, which is negotiated through the narrative construction process to accomplish social action. Complementing the focus on narratives and stories is the ethos building, partially evidenced through them. Within this overarching theory, I explore the concepts of credibility as well as character, especially as they relate to Edwards's ethos development, since he is the primary rhetor in this case study. I then situate a discussion of values, frames, and trust within the process of co-constructing ethos. The third major theoretical frame is environmental justice and its influence on narrative and ethos development, since all of the data gathered for this study revolve around one central question: what should be done to preserve Grey Cliffs so that everyone can continue using it? Exploring the answer to this question requires a developing rhetoric of relationship, a concept that permeates the theoretical framework.

Chapter 3, "Community Narratives and Ethos: Agency and Values," presents the need for some type of value alignment that needed to occur between the Corps and community before jointly accomplished social action could take place regarding the Grey Cliffs conflict. To demonstrate this need for alignment, I introduce values reflected in stories told by community members, such as values connected to religion, tradition, recreation, skepticism of government authority, and social unity. The value of social unity for this community also included the need for respect and the need for all voices to be heard. The stories these community members told strongly communicate the diverse voices involved and their potential to participate in resolving this conflict. Affective community values presented themselves in texts, such as narratives and stories, signs, fliers, newspaper ads, and social media communication, to begin to negotiate agency with the Corps to keep the area open.

Chapter 4, "Motivating the Compliant Individual: A Corps Resource Manager's Rhetoric of Regulation," focuses on David Edwards in his role as U.S. Army Corps of Engineers resource manager assigned to maintain and monitor the Grey Cliffs lake-access point. In this chapter that highlights Edwards's first attempt to persuade and engage with the community, I present Edwards as a regulator and motivator of action in this local community. Edwards saw the need for change to protect Grey Cliffs' sustainability; he also encouraged behavioral change in the community for the sake of public safety since crime had become so prominent in the area. But accomplishing social action was not easy for Edwards; while he attempted to create a persuasive persona, one that the community would accept, the community rejected him initially, based on his appeals to credibility alone. No meaningful relationship existed between Edwards and the Corps at this point. In the processes of communicating Corps values through various written, oral, and multimedia texts, Edwards presented values to the community to motivate it to comply with Corps rules and regulations. But this first attempt at persuasion met with strong community resistance.

"Attempting to Persuade as a Community Organizer: Norma's Narrative of Logic Without Emotion," Chapter 5, analyzes the role of Norma, the community organizer who coordinated the first community town hall meetings and motivated the community to attend them and who led the meetings to generate solutions to the problem of closing Grey Cliffs. Chapter 5 establishes Norma as a leader with a strong logos, or sense of logic, as she possessed strong organizational skills as well as experience with nonprofit organizations and grant writing obtained through previous work experience and volunteering. However, from the beginning of her involvement, Norma struggled with rumors about her circulating throughout the community and a damaged reputation. These rumors and her reputation threatened Norma's ethos with the community to the point that she was not able to rally the community behind her efforts; she was not able to co-construct an ethos with the community. Despite her strong logistical qualifications, Norma faced rumors focusing on three themes of her character: that she was overly controlling of the information, had a lack of personal connection with the community, and was untrustworthy. Despite these rumors, Norma initially constructed a persona regulated and motivated by her own apparent values of face-to-face communication, focusing on facts, rejection of emotion, consistency of organizational structure, and the importance of grassroots involvement. She also distributed several different types of texts, such as Convention of States literature, fliers advertising meeting locations and times, agenda and meeting notes, and researched rules for beginning nonprofit organizations. The values reflected in these documents and the documents themselves were helpful to the community as it started on its journey of negotiating some type of action with the Corps, but ultimately, the community resisted this regulation by deconstructing Norma's ethos and logos; Norma herself acknowledges her displaced agency as her role in the conflict resolution process diminished. This chapter demonstrates the need for a fully developed rhetorical persona necessary to persuade others to act; leaders require attention to their own credibility and assessment of community need, including responding to local culture and its receptivity rather than focusing on logic alone. This chapter also highlights the potential for deconstructed agency that results when a leader lacks established relationships and co-constructed ethos as prerequisites for rallying community members to act.

Chapter 6, "A Corps Resource Manager's Rhetoric of Relationship: Co-Constructing Ethos With a Community," analyzes ways that Edwards pivoted to address more relational concerns with the community based on an ethos of sincerity and affinity. Upon meeting resistance at first (see Chapter 4), Edwards consciously adapted his persona to portray a revised ethos so that the community would accept him and his message. To emphasize this change from highlighting regulations to attempting to foster relationships to achieve compliance, I analyze Edwards's reflective self-narrative using a rhetorical framework of character appeals. Ultimately, this new persona resonated effectively enough with the community so that social change could begin to be realized. To even begin this process, though, Edwards had to demonstrate willingness to co-construct ethos, agency, and an eventual new narrative with the community. For Edwards, a revised map, including negotiated plans for rejuvenating the area and materials the Corps would provide, served as one tangible way of building that trust.

Chapter 7, "Narratives of Jointly Accomplished Social Action Through Aligned Values: The Negotiated Resolution," argues that developing a co-constructed, common narrative through a revised framing strategy increased value alignment between the Corps and community in such a way that the community changed its behavior to arrive at a solution to this conflict that works for now. Analyzing semistructured interviews, I present the changed narratives, stories, texts, and actions of the community to identify specific ways the community responded to the ethos appeals Edwards extended in his conversations with the community. Texts the community constructed, such as fliers advertising cleanup days and social media posts communicating Corps regulations, verified its role in reflecting Edwards's reconstructed ethos back to him and agreeing to help the Corps with its rejuvenation and decriminalization efforts.

Chapter 7, then, indicates the results of efforts by Edwards and the community as they strived to co-construct an ethos that in turn fostered a new narrative that framed this conflict in a different, more positive, and more inclusive way than the "us versus them" type of narrative that the community promoted before. Together the Corps and community constructed a new narrative that contained values the Corps and community could agree upon, and a new value developed during this process: framing a positive future for Grey Cliffs. This new narrative and aligned values became evident to all who visited Grey Cliffs based on the results of these collaborative efforts. While not all values between the Corps and community would perfectly align, the Corps and community members found enough in common among their values to negotiate a workable solution to keeping Grey Cliffs open. These co-constructed narratives and ethos are a tentative representation and evidence of this jointly accomplished social action that could change at any time based on the changing, destabilizing activities surrounding Grey Cliffs.

Chapter 8, "The Continued Negotiation Process: Implications for the Future," contains my brief reflections on the current status of the negotiated resolution 3 years later. Keeping reflection at the forefront of this chapter, I discuss agency's cyclical nature and the ways Edwards and community members negotiated agency throughout this conflict as well as ways this co-constructed agency might continue to be renegotiated into the future based on changing values evidenced in narratives, ethos building, and social and environmental justice frames. This renegotiation process considers potential changes to interactions among Grey Cliffs, the Corps, and community as conditions change over time. I then present limitations of this work and suggest areas for future research. Toward the end of the chapter, I emphasize the need for continued reflection in conflict resolution contexts such as this one: rather than being a singular resolution event that participants negotiate once and for all, this conflict continues to be renegotiated over time, necessitating continued reflection about successes, potential improvements, and ways to incorporate additional participants who may become newly involved in the conflict resolution process. Finally, I present future implications suggested by the communication surrounding this conflict that include both global and local communication contexts.

My hope is that the case study in this book will prove useful to communicators from a variety of fields, as well as those influencing the creation of public policy, as they seek to learn more about how government representatives, for example, might connect with community members when trying to resolve sensitive issues. Organizational communicators, technical and professional communicators, and environmental science communicators will find these observations valuable when considering conflict resolution in general, especially conflicts involving diverse publics. In addition, scholars and students in these fields will find this work helpful; not only is this case study a model for ethnographic work others might want to replicate in the future, but the community-specific insights provide a window into the complex cultural, historical, and dynamic processes of engaging with community members. Through this work, readers learn more about ways rhetorical processes such as ethos negotiation provide avenues for increasing collaborative dialogue among diverse participants who bring their own cultural contexts to the negotiating table. Attempting to acknowledge voices and assign legitimation to all of these participants, this study should also encourage others to consider the broad communication dynamics that extend beyond the apparent simplicity of polarizing views that stand out at the beginning of crisis situations. While these polarizing views certainly require our attention, they often obscure the possibilities for negotiating ethos, relationships, agency, and possibilities for sincere persuasion that this book argues are essential for continued conflict resolution.

2
NARRATIVES, STORIES, ETHOS BUILDING, AND ENVIRONMENTAL JUSTICE

In many ways, the Grey Cliffs conflict contains similarities to other conflicts taking place in the fields of technical and professional communication, organizational communication, environmental science, and public policy: social actors and public participants on different sides of the fence attempt to persuade each other to accomplish their desired goals, often without using language the other side can understand and empathize with. As a result, neither side persuades the other to pay attention to their individual narratives. On one side of the Grey Cliffs conflict, Edwards represented the scientific, organizational perspective from the U.S. Army Corps of Engineers; he regulated the information flow about the Code of Federal Regulations, Grey Cliffs, and statistical information from the county sheriff's reports. On the opposing side, the community, while seemingly homogeneous in its unified response against Edwards, consisted of very different individuals with their own narratives, stories, goals, and ethos. These "competing rhetorics" (Druschke, 2018, p. 27), in this case, appeared to illustrate how Edwards, the "scientific" engineer, spoke "in abstractions and in analytic, unemotional ways" (Herndl et al., 2018, p. 72) while the community, similar to Herndl et al.'s (2018) discussion of farmers in their work, "talked about specific, concrete things and often about the affective values associated with issues" (p. 72). Yet these competing rhetorics also contain generative possibilities for solving conflicts, such as the ones the Grey Cliffs community experienced.

This chapter provides a theoretical framework used to analyze the data gathered about this conflict. First focusing on narratives and stories, I highlight the importance of symbolic capital, social capital, and agency in the process of negotiating and accomplishing social action through those narratives. The rhetorical framework of ethos furthers the analysis of data gathered during this fieldwork by clarifying the relationship (and sometimes contrast) between credibility and character during the ethos-building process. All parties involved in this conflict required co-constructed narratives and ethos to begin developing a rhetoric of relationship; negotiated values, value frames, and trust were all a part of that co-construction process. Finally, I situate the context and potential of these narratives, stories, and ethos building within environmental and social justice, since much of the communication surrounding the conflict includes the Grey Cliffs' nonhuman environment, including the embodied experiences of rural community members who spoke using affective valuations of this beloved place.

NARRATIVES, STORIES, COUNTERSTORIES, AND ANTENARRATIVES

Stories and narratives reflect organizational culture, and culture in general (Britt, 2006; Dunn, 2019; Faber, 2002; Heath, 1983), in memorable ways, and they also serve to construct an organization's identity, such as a government's. Counterstories often challenge accepted stories in order to accomplish change, and "antenarrative precedes the clean coherent narrative" (Small, 2017, p. 239) in fragmented ways that can ultimately influence dominant narratives as well. All of these play a part within evolving discourse surrounding a conflict and the potential relationships within that conflict.

Narratives

Within organizations, narratives serve to promote values and identities situated within organizational culture and history; these narratives can sometimes serve to control employees, for example, by ensuring employee attitudes and actions align with the overarching organizational narrative (Alvesson & Wilmott, 2002). Specifically, narratives can reveal themselves through messages an organization communicates, such as through slogans, rules and regulations, handbooks, oral messages presented during meetings, dress codes, scripts that employees use when communicating with outsiders, and other forms of written, oral, visual, and digital communication. Community narratives can reveal themselves through similar types of texts used to accomplish community goals as well as through stories of embodied experience that support community narratives and reveal personal and community identity. Narratives relate so closely to identity that *identity* can be defined as "the capacity to keep a particular narrative going" (Baumlin & Meyer, 2018; Giddens, 1991b, p. 54); narratives also closely relate to the process of *identification*, during which communicators emphasize similar values with their audiences (Burke, 1969; Walton, 2013). Jones and Walton (2018) indicate that narrative "is a promising tool for engaging explicitly with issues of diversity and social justice" (p. 243), in part "because of its capacities for fostering identification" (p. 243) as well as integrating "historicity" within local contexts. The concept of "historicity, which posits that ideals, values, and beliefs are developed, articulated, and rearticulated through perceptions of historical and contextual occurrences" (Jones & Walton, 2018, p. 251), emphasizes "cultural context and representation" (Jones & Walton, 2018, p. 251). Historicity, therefore, illustrates the ways an organization's and community's cultural histories, for example, influence the communication of values and identities within developing narratives.

The Grey Cliffs narratives presented in this case study are some of the texts, broadly defined and appearing through various types of media, that reveal the cultural-historical experiences and values of this local rural population as well as those of an organizational representative, Edwards, during this dialogic communication process. During conflicts, narratives, values, and identities can clash when different narratives appear to lack potential for alignment. Understandably, communicators demonstrate investment in particular narratives because they have adopted the values they promote, and, as a result, they feel their own identities are connected to these narratives and oftentimes feel defensive in protecting them. In this case, in order for the community and the Corps to begin co-constructing a common narrative, the old, dominant, polarizing narratives of the environmental degradation at Grey Cliffs, the historical narrative of Corps control, and the relationship between the Corps and community needed to change in such a way that the community and Corps could work together to co-construct a shared narrative, conveying a joint rhetoric of relationship, that all could agree with to some degree.

Stories

Here I distinguish between narratives and stories similar to how Jones and Walton (2018) do in their work connecting narratives and social/environmental justice. Drawing upon Arnett (2002), Jones and Walton (2018, p. 243) clarify that *narratives* function as means of participation, and, within those, *stories*, more narrowly defined, contain a plot structure and relate more specifically to a storyteller. At times, particularly during times of change, stories might also be fragmented and less defined in their structure (Boje, 2001, 2008, 2015; Small, 2017; Syed & Boje, 2015). Narratives contain stories but also emphasize participation and context, defined by those involved.

Stories are very personal and reveal much about values and individuals as well as collective identity (Baniya & Chen, 2021, p. 79; Small, 2017, p. 240), since "stories—oral, written, and visual—compose a tapestry that represents a person's life" (Mangum, 2021, p. 57) and living together as a community. Because of these strong personal connections, stories can also be influenced by "intuition" (Sackey, 2018, p. 150), which "encompasses the everyday experiences situated in local contexts and the values that stem from experiences within those spaces" (p. 150). Further, according to Sackey (2018), "intuition, as a form of contextual reasoning, when coupled with personal stories and empirical data" (p. 151), allows participants to make connections between themselves and the environment surrounding them, making stories a key to others' understanding of participant values and participants' nuanced understanding of how their own everyday experiences and values relate to the environment.

As an organizational representative, Edwards bolstered his discussion of rules and regulations with stories at the first town hall meeting. In this case, the stories Edwards told of his experience at Grey Cliffs served to document his concerns and justify his approaching the community members from his identified and assigned position of authority. As Faber (2002) writes,

> Put into the context of organizations, identity-stories provide agents within organizations ways to define themselves within and against cultural, historical, social, organizational, and economic structures. Identity-stories are powerful because they interpret perceptions of everyday life. They become filters through which people interact with the world. . . . An organization can be so focused on its own identity-story that it ignores the identity-stories competitors, clients, suppliers, and customers are telling. (p. 171)

In times of conflict such as this, Edwards was interested in listening to community stories, but he was more interested in ensuring the illegal activities at Grey Cliffs stopped. His "identity-story" at this point was all about compliance in order to achieve his organizational goal, rather than about developing community relationships.

In this case study, in addition to Edwards's stories that initially originated from an organizational perspective that focused on the transgression of regulations, the community's cultural-historical stories played a huge role in the Grey Cliffs conflict. These individual stories documented narratives of loss from the community's experience. These stories also conveyed affective community values about how much community members enjoyed Grey Cliffs as a physical place, including stories of off-roading that damaged the landscape environmentally. Importantly, these stories revealed relationships that community members had not only with each other but with Grey Cliffs as a geographic space. These types of stories unified the community. In addition to these types of relationship-building stories, communicative agents can also challenge dominant stories through counterstory.

Counterstories

Counterstories (Delgado, 1989; Dunn, 2019; Martinez, 2020; Rea, 2021) "are stories from the margins that disrupt and challenge dominant cultural narratives" (Dunn, 2019, p. 20). Martinez (2020) defines them as "empower[ing] the minoritized through the formation of stories that disrupt the erasures embedded in standardized methodologies" (p. 3). Counterstories that contribute to alternative narratives also provide agency for those creating and participating within them (Butts & Jones, 2021, p. 14; Martinez, 2020), especially since they highlight the "centrality of experiential knowledge" (Martinez, 2020, p. 9), a tenet Martinez identifies with critical race theory. For communities such as the one surrounding Grey Cliffs, this type of story plays a particularly important role; at first, counterstories were the only way community members could interact with the conflict taking place around them, the only way they could object to the dominant Corps narrative and the sheriff's report statistics. These counterstories revealed embodied knowledge these residents associated with their identities as longtime participants in Grey Cliffs' activities. While the concept of counterstory can often highlight agency within racially and ethnically marginalized groups, for example, the Grey Cliffs community did not physically *appear* to be racially marginalized, although many residents of this area proudly claim some degree of Cherokee heritage, since the Cherokee people resided in this area centuries ago. However, the concept of *counterstory* can also apply to "minority groups of socioeconomic class" (Jones et al., 2016, p. 215), and the Grey Cliffs community, as a rural population, falls into this category to various degrees, as a predominantly working-class community.

Antenarratives

Antenarrative highlights the fragmented, unpredictable nature of discourse that contrasts with a more orderly view of standardized *narrative*, those discourses that are already accepted within organizations (Boje, 2001, 2015). As Small (2017) emphasizes, "Antenarrative theory and method embraces such uncertainty and 'mess' in research in communication" (p. 242), such as the Grey Cliffs conflict that includes different types of groups and stakeholders with seemingly different values and perspectives.

Boje (2015) characterizes antenarrative as "a prospective sensemaking (looking forward), and is an intraplay with now-spective (in the present moment of emergent being), and retrospective (backward looking) manners of sensemaking. The agential aspects of antenarrative are in its intraplay with materiality" (p. 1). Antenarratives also include stories that reveal embodied experiences and localized meanings (Small, 2017). As Jones et al. (2016) describe antenarratives, "They link the static dominant narrative of the past with the dynamic 'lived story' of the present to enable reflective (past oriented) and prospective (future oriented) sense making (Boje, 2008, pp. 6, 13)" (p. 212). Key to antenarrative is the interaction of past, present, and future sensemaking. Because "antenarratives morph and coalesce in storytelling networks" (Boje, 2015, p. 1), stories from communities such as Grey Cliffs are key to making sense of a past, present, and future that may seem very different and disjointed. Like counterstory, antenarrative is seen as agential because of its dialogic nature and focus on social relationships (Boje, 2015), as well as the amplification of individual voices within those relationships and local contexts (Mangum, 2021). This type of antenarrative and storytelling introduces "fragmented living stories of people situated in unequal power relationships in organizations" (Syed & Boje, 2015, p. 48) that would otherwise not be present within more stable, linear, accepted organizational narratives. The living, dynamic nature of antenarrative contrasts with the apparent stability of the accepted, organizational narrative that can often be marginalizing; these living stories "highlight the fact that the character of interpretations and experiences is always open, polyphonic, equivocal, dialogical, unfinished, and unresolved" (Jørgensen, 2015, p. 284). During parts of their interaction with the Corps, this community's access to and use of their own living counterstories and antenarratives were the only ways their voices could be heard.

While in this chapter I define narratives, stories, counterstories, and antenarratives as part of the theoretical framework of this case study, I want to emphasize that these are not essentialized, non-overlapping categories. For example, stories are a part of antenarrative (Small, 2017) as well as narrative, and counterstories could appear within narratives and antenarratives as well. In addition, parts of antenarrative could evolve to become part of dominant narratives, and more dominant, accepted narratives could undergo a regression process that relegates them to positions of decreased dominance, based on different forms of sensemaking.

Stories, counterstories, narratives, and antenarratives could present themselves in various stages of development, as they function to mediate different types of social action and relationships, which are also in a state of change. Within a particular cultural context, narratives constantly undergo revision and negotiation, stabilizing and destabilizing in a constant state of flux, based on similarly varying degrees of stabilizing and destabilizing power relations. On both the Corps' and community's sides, these various forms of narratives and stories demonstrate the relational processes of negotiating symbolic capital, social capital, and agency; in other words, these narratives and stories were poised for opportunities as potential discourses of power and social change.

Narratives and Stories as Symbolic and Social Capital

Defined as "accumulated labor . . . which, when appropriated on a private, i.e., exclusive, basis by agents or groups of agents, enables them to appropriate social energy in the form of reified or living labor" (Bourdieu, 1986, p. 241), Bourdieu's (1986) concept of *capital* complements the role of Edwards's and the community's narratives, antenarratives, and stories; each side worked to negotiate the potential for power and agency among those involved in this conflict surrounding Grey Cliffs. In this case, the distribution of capital through narrative enabled action for some community members while constraining others. Bourdieu (1986) writes,

> And the structure of the distribution of the different types and subtypes of capital at a given moment in time represents the immanent structure of the social world, i.e., the set of constraints, inscribed in the very reality of that world, which govern its functioning in a durable way, determining the chances of success for practices. (p. 242)

This enabling and constraining represents the destabilizing "interactionality" (Chávez, 2013, p. 58; Walton et al., 2019, p. 126) that often occurs and recurs within power structures during the negotiation process. At first, even though community members were free to speak and promote their own narratives and stories at the first town hall meeting, these potential social actors' voices were constrained in their "chances of success for practices" (Bourdieu, 1986, p. 242); attempts to accomplish their intended, negotiated goals contributed to a "polyphony" of voices (Bondi & Yu, 2019) as each expressed individual roles in participating in the solutions proposed or objecting to them, albeit chaotically. These voices were also constrained at the time because of a lack of authority to legitimate the various discourses they were voicing.

In contrast to Edwards's cultural capital that was communicated by Edwards as a singular Corps representative from the scientific, governmental, enforcement side, the community possessed a different kind of capital—a strong social network. According to Faber (2002),

> certain symbols—manners, dress codes, automobile styles, addresses, vacation spots, language use, and pronunciation—stand in for the use of capital. Bourdieu refers to this as symbolic capital and proposes that researchers consider the ways in which groups use symbolic capital as social strategy: to differentiate themselves, select members, and wield power. (p. 156)

Capital can't really be used, though, if the "other" doesn't recognize and value that capital. That disconnect explains some of the conflict at the beginning of the interactions between Edwards and the community: because Edwards hadn't heard the cultural-historical stories of the area from individual community members, he was not prepared with a similar narrative or stories to generate goodwill with these members. Because "narratives denote what . . . Bourdieu calls the symbolic capital of the community" (Faber, 2002, p. 33), the community, while differing in their individual narratives, stories, and rationales for why the area was important to them, needed to construct some sort of narrative with a common theme that would resonate with Edwards; the community needed an effective counterframe before it could even begin co-constructing a new narrative frame, a rhetoric of relationship, with Edwards. As "agents struggling to acquire these forms of capital" (Schryer et al., 2007, p. 25), the community underwent hardship and the "unfortunate part" that Edwards referred to regarding the community experience when he first mentioned closure as an option at the first town hall meeting. Edwards regretted that this initial proposal to close the area distanced himself even more from the community; he already possessed the symbolic/social capital assigned to him as a government representative, and the community, not realizing how powerful its narratives and stories were, felt powerless and rebelled before even attempting to negotiate.

Eventually, Edwards decided to extend his narrative of sincerity, affinity, and credibility in such a way as to engender trust with the community in an effort to increase his growing symbolic, social capital; as a result, he began making "symbolic *investments*" (Bourdieu, 2007, p. 180) in the community through his efforts, narratives, and compromises to the solutions proposed to Grey Cliffs' problems. Introducing these new narratives as social capital was no easy task, since "the members of the group must regulate the conditions of access to the right to declare oneself a member of the group" (Bourdieu, 1986, p. 251), and this group was an exclusive one made up of insiders. Exchanging social capital necessarily involves relationships, ones that involve "an apparently gratuitous expenditure of time, attention, care, [and] concern" (Bourdieu, 1986, p. 253). Regarding this type of capital exchange and environmental issues, Martinez (2022) offers as well that effective "environmental communication . . . focuses on relationships and incorporates two-way communication that addresses the values of readers on a scale they comprehend" (p. 176); narratives and stories in their various forms reveal values that must be shared during the co-construction of knowledge and action.

Truly, these narrative exchanges and co-constructions were symbolic because there was no way to guarantee how long these shared narratives would complement authentically changed actions, how long this exchange of social capital would continue working. But clearly "a common narrative was needed" (Faber, 2002, p. 8), and narratives and stories played a defined role in discussing, debating, and assigning action to clean up the Grey Cliffs area, which in turn entailed the process of negotiating agency.

Narratives and Stories as Tools for Negotiating Agency

The polarized Grey Cliffs communication scenario resulted in a stalemate at first because Edwards appeared to have all of the authoritative, regulatory, and discursive power, and the community had no idea how to begin interacting with Edwards to accomplish what it wanted. This conflict illustrates the power differential between Edwards as the proponent of technocratic discourse (a macro view) and the community, who conveyed their expertise through personal, embodied experience as storytellers (a micro view [Walton et al., 2019, p. 106]). Moreover, Edwards and the community had no real relationship with each other, so no conversation inroads even existed. In order to move forward in a way that could benefit the Corps mission, community values, and the Grey Cliffs environment, the Corps and community would need to co-construct and negotiate agency through these narratives and stories. Herndl and Licona (2007) define agency as "the conjunction of a set of social and subjective relations that constitute the possibility of action" (p. 135), and this conjunction necessarily involves meaningful relationships among the involved participants. As Grabill et al. (2018) write, "To . . . understand rhetorical agencies requires attending to the construction of relationships between humans and nonhumans, built and natural environments, history/archives and the present" (p. 195). Attending to these complex relational dynamics, as George and Manzo

(2022) emphasize, is a prerequisite to addressing environmental issues; doing this "ensures respect for the agency of community members and the culture of the community" (p. 107).

Emphasizing the tenuous nature of this agency-negotiation process, Herndl and Licona (2007) state, "It [agency] does not reside in a set of objective rhetorical abilities of a rhetor, or even her past accomplishments. Rather, agency exists at the intersection of a network of semiotic, material, and yes, intentional elements and relational practices" (p. 137). Connecting the concept of agency to Giddens and power as well, Herndl and Licona (2007) highlight the "truly social phenomenon" (p. 137) that agency is. These scholars note that this type of community decision process about what values contribute to the common narrative allows "nondominant groups" (p. 143) to participate in the power and agency negotiation process (pp. 142–143): "Agency is the name we give to this rearticulation of cultural rhetoric" (Herndl & Licona, 2007, p. 150).

To clarify further, this agency negotiation process entails recognizing and being constrained by genres and social structures, reinventing them for social purposes, and constructing narratives and stories based on these processes. Working on such a process can be very difficult for some community members, though, as we see in the Grey Cliffs conflict: Edwards, the representative of "authoritative practices" (Herndl & Licona, 2007, p. 143), certainly could have prevented the community from participating in any kind of agency negotiation during this process. His values that included hearing the public voice when making decisions about Corps lands and promoting a more positive view of the Corps, though, motivated Edwards to negotiate agency with the community to build a collaborative solution. Agency negotiation options for the community were complicated, though. According to Faber (2002), "agents must choose to replicate social relations in order to build and maintain power" (p. 120). While the community did negotiate agency to varying degrees throughout this process by asserting various voices and stories, Edwards's power as Corps resource manager would inevitably remain a priority. Edwards and the community, therefore, would need to participate in a continual power-negotiation process to enable this precarious collaboration to continue taking place, and the community would need to continue reproducing the necessary social relations (including evidence of compliance) in order to ensure continued access while at the same time gaining empowerment to act.

This back-and-forth interaction between Edwards and the community indicates continual constraining and enabling agency among all parties. As Herndl and Licona (2007) write,

> We . . . also suggest that authority and agency are not always opposing forces within complex institutions. We need a more careful understanding of the interaction between agency and those regulative forces that stabilize institutions and practices. Indeed the regulative power of rhetorical and institutional authority is often interrupted as agentive and authoritative motives or, to use Burke's term, overlap. (p. 134)

Edwards, as a Corps representative, and the community illustrate this type of overlap, since everyone worked closely together to negotiate and co-construct agency in a way that certainly is not perfect.

Ultimately, agency, ethos, and authority are connected through the process of negotiating social action:

> Like agency, authority is a social location, (re)produced by a set of relational practices. The authority to speak—a speaker's authority in discourses and debates—is a social identity that is occupied by a concrete individual but emerges from a set of social practices. In this sense, authority is tied to classical notions of ethos. (Herndl & Licona, 2007, p. 142)

Agency hinges on meaningful relationships, and participants use narratives and stories as tools for social capital to negotiate agency and work within established power structures (Giddens, 1984, 1991a, 1991b), creating adaptations that can work to support collaborative action for change. But, importantly, this process involves not simply uncritically replicating established power structures but redressing inequalities, lack of diversity, and lack of inclusion within these structures, as social justice advocates promote (Walton et al., 2019).

Because narratives and stories have such a key role in this dynamic process, the second main part of this study's theoretical framework is the rhetorical concept of ethos. In order for any type of resolution to move forward, Edwards would need to construct an ethos through his organizational narrative as a Corps representative that community members would be willing to negotiate with, and community members, individually and collectively, would need to demonstrate an ethos through their narratives and stories that would persuade Edwards. An evolving rhetoric of relationship between Edwards and the community could be one possible result of negotiating ethos. Only within the context of these relationships would these social actors be willing and able to move past the polarization that initially characterized this conflict. Because narratives often contain stories, references to *narrative* throughout this case study imply that stories could play a part within these narratives, as well. Specifically, I include community stories to indicate ways counterstories and antenarratives are part of negotiating ethos and relationships with

Edwards as a Corps representative, all contributing to a revised, developing, new narrative of Grey Cliffs as a physical space.

ETHOS BUILDING

Rhetorical analysis based on Aristotelian rhetoric (Aristotle, ca. 367–347, 335–323 B.C.E./2019) is essential to explore and uncover rhetorical strategies participants use to interact with various audiences such as those involved in the Grey Cliffs conflict. Discourse analysis is inherently important, as well, as it relates to audiences of and purposes for the discourse (Allen et al., 2012, p. 213; Gee, 2014; Hyland, 2012; Paltridge, 2012; Pollach, 2018). These discursive, narrative perspectives reveal the agency negotiation and social capital exchange process working between Edwards and the community as they created more reciprocal unity through aligned sustainability values to keep Grey Cliffs open. Rather than representing an essentialized unity, though, this fluid, dynamic relationship indicates the need for continuing, reflective, rhetorical persuasion through dialogic communication.

Below, I contextualize the discussion of ethos by dividing it into two parts: credibility and character. I then connect ethos to the relationship-development process before moving to clarifying the connections among narratives, ethos, and accomplishing social action. Establishing ethos through narratives/stories to accomplish social action includes dynamic processes of aligning values, co-constructing value frames, and negotiating trust.

Constructing Ethos Through Credibility and Character Development

Simply defined, ethos is the credibility of the speaker (Aristotle, ca. 367–347, 335–323 B.C.E./2019). However, the concept of ethos presents a lot more complexity when we consider an outsider such as Edwards and his task of constructing a credible ethos to persuade a community of "insiders" to preserve Grey Cliffs. In Book I of his *Rhetoric*, Aristotle clarifies that credibility, such as appeals to authority, expertise, and experience, is enhanced by character: "Persuasion is achieved by the speaker's personal character when the speech is so spoken as to make us think him [sic] credible. We believe good men [sic] more fully and more readily than others" (Aristotle, ca. 367–347, 335–323 B.C.E./1990, p. 153). As Aristotle discusses a speaker's character in more detail in Book II of the *Rhetoric*, he clarifies that "good sense, good moral character, and goodwill" (Aristotle, ca. 367–347, 335–323 B.C.E./1990, p. 161) will "inspire trust" (p. 161) in the audience. Since each of these qualities requires goodness, the question remains how rhetors might demonstrate or co-construct an ethos of goodness for and with their audience. Goodness can be accomplished through developing character virtues, which Aristotle discusses further in his *Nichomachean Ethics* (Aristotle, ca. 350 B.C.E./2012).

Contemporary scholars have identified and adapted some of these virtues as they have discussed various dynamics of ethos and developing relationships with audiences. For example, Baumlin and Meyer (2018) discuss the process of a rhetor's constructing an inner self as well as an outward presentation of that self; during this process, the rhetor hopes to display "sincerity, authenticity, and self-consistency" (p. 6). Braet (1992) clarifies that ethos results from the interaction that a speaker has when communicating with an audience; the audience assigns ethos to the speaker, and the speaker "argues" based on that assigned ethos, and vice versa (p. 311). Because the outward presentation of the self is so important, rhetors must also display an ethos that aligns with the audience's culture; only then will the audience view the rhetor as having a common and believable connection with it based on outward actions and communication. As Campbell et al. (2015) state, ethos refers "to the ways in which you mirror the characteristics idealized by your culture or group" (p. 251). Likewise, Halloran (1982) specifically defines ethos as "manifest[ing] the virtues most valued by the culture to and for which one speaks" (p. 60). This "similitude," "the appeal to similarities between the author of the text/speaker and the audience" (Higgins & Walker, 2012, p. 197), "perceived similarity" (McCormack, 2014, p. 139), "or liking between him [*sic*] and [the] audience" (McCormack, 2014, p. 139) is just one analytical category that some researchers have applied to various types of texts to learn more about communicators' persuasive techniques regarding differing value orientations, as well as values and social/environmental issues, such as those occurring at Grey Cliffs.

Establishing similarity between the speaker and audience is crucial and can be achieved through various types of communication strategies, such as using "familiar, colloquial language, . . . relaxed body language, open and honest facial expressions, and a friendly . . . tone of voice" (McCormack, 2014, p. 139). According to Braet (1992), when the rhetor and audience's "preferences" (p. 313), or values, coincide, the audience then views the rhetor as good and therefore trustworthy (p. 313). Demonstrating this alignment can be difficult, though, since it relies on the rhetor's *estimation* of the values the audience holds (Halloran, 1982, p. 63). In Edwards's case, he needed to develop an "invented ethos" (Mackiewicz, 2010, p. 408) for the specific rhetorical situation of the first community town hall meeting, when he began to address this conflict publicly, as well as a "situated ethos" (Mackiewicz, 2010, p. 408) that indicated credibility over time; this type of credibility needed to be connected specifically to community values and Edwards's character in order for him to persuade community members effectively.

Connecting Ethos and Relationship

Because language is a form of social action, developing relationships with audience members is one way to co-construct ethos and shared identities (Cheng, 2012, p. 427), especially through rhetoric (Mackiewicz, 2010); the rhetor and audience are engaged in accomplishing the same tasks, and all of these actions are governed by shared language use. Necessarily, character and ethos play a role in this discursive construction when the rhetor emphasizes affinity with the audience; in other words, "I am like you, and together, we value the same things." These "source relational attributes" (McCormack, 2014, p. 138; Weresh, 2012, p. 234) continue to bestow favorable, "friendly" characteristics on the rhetor that make the communicator appealing to the audience. Feedback on these characteristics from audience members (Mackiewicz, 2010) contributes to constructing a communicator's credibility and, in some cases, expertise (Mackiewicz, 2010); the combination of credibility and character can lend additional persuasiveness to rhetors as they connect with their audiences, and vice versa. In fact, these relationships are part of the rhetor's and audience's continual revisiting of the ethos construction process, as different parts of it are "negotiated and reshaped" (Mackiewicz, 2010, p. 421) through communication. An example of this occurring could be when an audience accepts a rhetor's expertise once it learns more about the speaker's character through meaningful relationship and interactions. More specifically, as George and Manzo (2022) write, "Respect and active listening, sharing meals, getting to know elders are all critical components to building successful communications" (p. 107). These social interactions contribute to the narrative and ethos co-construction process.

Co-Constructing Narratives and Ethos to Accomplish Social Action

Co-constructing narratives and ethos through these relationships is a prerequisite for collaborative social action. While this complex process can occur in a variety of ways between the organizations and the publics and communities they communicate with and serve, here, I focus on three parts of the narrative and ethos co-construction process: aligning values, co-constructing value frames, and negotiating trust. All of these elements need to have some degree of similarity for participants to build joint narratives and ethos to accomplish common goals.

Aligning Values

Commonly, organizations such as the Corps and communities such as these operate from different values and value systems, and, as a result, they experience contradictions that can inhibit accomplishing goals together. Often, organizational values differ from those of local communities because "everyday experiences situated in local contexts and the values that stem from experiences within those spaces" (Sackey, 2018, p. 150) often differ from more removed, organizational values; those values appear separate from the "mundane and everyday experiences" (Petersen, 2018, p. 5) that communities build to create important "experiential knowledge" (Petersen, 2018, p. 5). For example, as Patterson and Lee (1997) point out when discussing the regulatory discourse surrounding the Kingsley Dam relicensure, governmental, organizational language focused on the "technical language of functionalism" (p. 29) while the public "[spoke] in anecdotes about the virtues of a way of life" (p. 29), similar to observations other scholars have made about scientists speaking in abstractions while local publics draw upon affective values connected with specific, embodied experiences with the environment (Herndl et al., 2018, pp. 68–69, p. 72; Peterson, 1997; Sackey, 2018). In these cases, values founded in science clashed with those of everyday, lived experiences.

Scholars indicate that in order for organizations and the communities they serve to accomplish common, desired goals, they must have similar value orientations (Lehtimäki et al., 2011; Martinez, 2022, p. 175). In the case of environmental communication, organizations can connect sustainability to a community's values based on specific places that they value (Martinez, 2022, p. 175; Schweizer et al., 2013). Connecting to specific, physical, meaningful places also taps into "embodied knowledge" (Sauer, 2022) critical for community members' understandings of sustainability values, as these values apply to the geographic places that people experience individually and locally. For these reasons, focusing on "local contexts" is a "common theme that resurfaces repeatedly in both scholarship and environmental discourse" (Tillery, 2019, p. 76). Sometimes such values between organizations and local communities/individuals may not be that different but are expressed in different ways, such as through the polyphony (Bakhtin, 1984; Bondi & Yu, 2019; Castelló et al., 2013; Christensen et al., 2011) of voices that the community was expressing during the first community town hall meeting that communicated various community values. These differences can make it difficult for organizations to even identify public and community values that need to be aligned with organizational ones, and vice versa. Another reason for this difficulty is that there is no one way to define the public due to the polyphony of local voices (Bakhtin, 1984; Bondi & Yu, 2019; Castelló et al., 2013; Christensen et al., 2011) participating in the social construction of discourse. This concept of polyphony in organizational communication highlights the need to listen to various stakeholders' voices, recognize their diversity, and attend to various perspectives when making decisions, similar to what Le Rouge and Stinson (2022) and George and Manzo (2022) argue specifically regarding environmental science communication.

From a specific organizational rhetor's perspective, in the process of developing and negotiating an ethos that addresses an audience's values, a rhetor essentially needs to co-construct shared *identities* with the audience. These shared identities are based on dedication to the same cultural values as well as trust and respect that the audience develops as it communicates reciprocally with the rhetor, generating "positive recognition" (Martínez, 2012, pp. 3926–3927) among the participants. In this case, Edwards attempted to develop a shared identity with the community as he motivated it to value what he and the Corps valued: safety and environmental protection of public lands.

The Burkean concept of identification (Burke, 1969) "overcomes division and unites individuals along various possible lines of interest (beliefs, motives, tastes, etc.) setting the stage for persuasion" (Cheng, 2012, p. 426). As a rhetor, "identifying" (Burke, 1969, p. 55) with an audience through various discursive strategies that emphasize these similar values is essential for the persuasion process. According to Walton (2013), "Burke extended Aristotle's conception of rhetoric with 'identification,' claiming that an effective way to persuade people to a particular belief is to show a connection to their other beliefs and values, to identify with people" (p. 89). Beliefs and motives coincide with values: our values help us construct beliefs conveyed through individual and group action, and our values motivate us to accomplish action. (I acknowledge here that Burke's antisemitism gives scholars pause, such as Fernheimer [2016], who cautions us to "trouble the way we rhetoric scholars so easily and uncritically embrace the theoretical apparatus he developed under such deeply entrenched anti-Jewish attitudes" [p. 52].)

A prerequisite to developing shared identities through a rhetoric of relationship is not only developing shared values but also communicating about them in meaningful ways that will resonate with the intended audience (Heracleous & Klaering, 2014, pp. 134–135), whether that audience is made up of community members, groups, or organizational members. Not doing so results in disorientation (Anson & Forsberg, 2002; Gaitens, 2000; Kohn, 2015, p. 176) and dis-identification (Alvesson et al., 2008, p. 19), defined as "constitut[ing] the self around what it is not." As participants perform identity work with other social participants to negotiate action and power, "identity struggle" often occurs, which is "a source of tension and pain, as it is at odds with the new organizational regime and makes smooth adaptation and compliance more difficult" (Sveningsson & Alvesson, 2003, p. 1186). One way to address contradictions in the relationship-building process and facilitate value alignment is through co-constructing value frames.

Co-Constructing Value Frames

The ethos-building process can also include co-constructing value frames that help create a new narrative around the original conflict, framing the situation more positively through the help of as many participants as possible. Clearly, at the beginning of this conflict, Edwards and the community communicated from vastly different value frames that prohibited aligning their actions about Grey Cliffs. Smith et al. (2020) indicate that

> if frames are the resource, or outcome, present in a social situation, then it is crucial to look at the activity of framing that enacts those frames—especially if those frames help individuals and organizations to reach desired goals. (p. 6)

Smith et al. (2020) define the process of framing as "socially legitimizing a message or way of thinking" (p. 3). In the past, the process of constructing frames has been viewed as the

> purview of leaders and managers [alone]. Less is understood about how workers within an organization may frame issues. For example, research on failed change efforts attributes unsuccessful transformation to a failure of leaders to craft a clear and compelling vision statement that directs, aligns, and inspires actions. (Smith et al., 2020, p. 3)

Organizational leaders often attempt to motivate changes in communities, too, and failure to accomplish the desired action often occurs because other participants' value frames need to be considered and addressed as well; otherwise, one dominant frame may simply be replaced by another without dialogue and compromise.

As Waller and Conaway (2011) discuss, "Frames function to emphasize the importance of an issue, to promote a specific interpretation of an issue in terms of causal factors, to introduce evaluative judgments on the parties to an issue, and to promote specific remedial action" (p. 87). This research and others discuss frames as constructed *against* something, whether that might be social injustices or another type of power structure (p. 88; see Cornelissen & Werner, [2014]; Waller & Conaway, [2011]; and Waller & Iluzada, [2020] for thorough literature reviews on this type of framing; also see Goffman, [1974]). While this frame-construction process sounds somewhat deterministic, judgmental, and individualistic, the language choices rhetors make allow framing to contribute to the process of co-constructing a *shared* narrative, a process that in turn opens up opportunities for participants to negotiate agency for action as they reframe conflict together. This type of framing strategy can have a significant impact on ways community members understand and perceive sustainability issues (Eubanks, 2015, pp. 124–131), especially since "paying attention to framing is essential to our becoming aware of the values that we genuinely hold" (Eubanks, 2015, p. 129). For technical and scientific communicators, choosing frames that resonate well with their intended audiences is critical to persuading and reaching these audiences (Smith & van Ierland, 2018), since differing values and communication strategies between engineers, for example, and the public often complicate constructing common value frames. Scientific and environmental communicators' intentional framing, based on audience awareness, can increase public participation and interest.

Framing clarifies values, and, often, environmental communicators will demonstrate these values through narratives in order to persuade community members to align their values toward particular environmental communication efforts. For example, scholars such as Ross (2017), Tillery (2017), and Walsh (2010) discuss how analyzing environmental communication through the lens of rhetorical *topoi* and commonplaces reveals recurrent themes in these narratives, such as "the scientist as hero" and "science as hubristic enterprise" (Tillery, 2017, p. 45). These commonplaces resonate with audiences who may subconsciously recognize these argumentative frameworks and may understand the environmental issues at hand better as a result. While environmental communicators can't ensure persuasive success by integrating these persuasive cultural frameworks into their discourse (Ross, 2013), these commonplaces and "narrative frames" (Tillery, 2017, p. 55) can serve as a way to introduce values to audiences in a way they might recognize and then accept, essentially aiding in the ethos-negotiation process by fostering the co-construction of meaning. In fact, evoking these shared cultural and value frameworks may even encourage trust in rhetors because "they embed values about what types of knowledge are good or moral and what kinds of institutions are trustworthy" (Tillery, 2017, p. 45). In emphasizing social justice value frames, some scholars (Walsh & Boyle, 2017) advocate a topological approach in moving "*beyond* intervention *to* [rhetorical] invention" (p. 2, emphasis in original) to "compose—not just deconstruct—political dynamics" (p. 4), using post-critical rhetorics, such as Lacanian topologies (Cowan, 2017). These rhetorical topologies encourage inventional possibilities for change beyond seemingly insurmountable power structures, providing even more opportunities to address opposing value frames as well as opportunities for negotiating trust.

Negotiating Trust

Based on Aristotle's *Rhetoric*, we trust speakers that demonstrate the characteristics of "wisdom, virtue, and goodwill" (Aristotle, ca. 367–347, 335–323 B.C.E./2007, p. 112; see also Tillery, 2006). Similarly, Griffin (2009) states, "To create this ethos of confidence in his character, the speaker must demonstrate good sense, good moral character, and goodwill (p. 1378a). Doing so, says Aristotle, will inspire trust in the speaker and therefore trust in the message" (p. 64). Part of demonstrating these character virtues is establishing a relationship with the audience, a prerequisite for displaying these virtues. As Weresh (2012) concisely states, "Relationship persuades" (p. 229). Complementing the emphasis on the importance of these character elements, Griffin (2009) discusses the pharmaceutical company Merck and how, based on problems with the medication Vioxx, the company needed to address a "fundamental contradiction in its relationship with the public" (p. 68) that included a lack of trust. In order to address this contradiction, Merck needed also to address primary narratives surrounding the problems with the medication, such as the stories surrounding Merck's responsibility in this situation: the conflict could appear to be unintentional or intentional, based on different views of the events. Merck had to address these contradictory narratives in order to facilitate trust with the public again; in essence, it had to create a shared narrative to regain and maintain trust (Weresh, 2012).

Similarly, Isaksson and Jørgensen (2010) connect trust with ethos and subdivide trustworthiness into "integrity, justice, truthfulness, courage, and passion" (p. 130). As they elaborate on trustworthiness, they state:

> We decided to use the term of empathy synonymously with goodwill/ perceived caring as we think it stresses the emotional and interpersonal qualities of the writer instead of his or her more consciously planned and

strategic intentions. Expertise, trustworthiness, and empathy each capture a quality that is practically internalized in the writer, whereas goodwill suggests an adopted stance of the writer. (Isaksson & Jørgensen, 2010, p. 132)

Because "empathy is thus relationship oriented" (p. 133) as well, a communicator needs to attend not only to his/her achievements (Isaksson & Jørgensen, 2010, p. 133) but to the concerns of the audience, which can be addressed by displaying a "you attitude," rather than a "me attitude" (p. 133). Creating this shared identity with the audience (moving from "me" to "you" or "we") also engenders more trust with an intended audience, as the communicator appears to be "one of them," an insider who is trying to accomplish goals that are important to the audience. In this sense, as communicators reciprocally develop trust, identity creation also develops as a shared, negotiated process between communicator and audience (Mackiewicz, 2010; Shamir et al., 1998; Weresh, 2012). Olman and DeVasto (2020) also demonstrate the importance of this type of reciprocal trust development as they discuss the need to "link experts and nonexperts in resilient configurations" when creating environmental risk visualizations (p. 20) "that outline a new way of thinking about 'effective' risk visualization as focused not on the visualization in question or an individual's ability to comprehend it but on the way that visualization gathers, solidifies, and equips political collectives to manage Anthropocene risk" (Olman & DeVasto, 2020, p. 24). Olman and DeVasto's work emphasizes that trust development goes beyond more "traditional" genres of oral and written technical communication to visual communication as well.

Importantly, in order for trust to be negotiated, influential community members need to be on board and convinced of the beneficial relationship with the communicator; these interventions create more trust among community members if the relationships are "local and interpersonal" (Mooney et al., 2022, p. 126); this type of trust is essential among community members to motivate them to participate in positive environmental efforts, as Mooney et al. (2022) found in their work on private well water contamination (p. 138). Scholars note that addressing "human embodiment" and relationships (Le Rouge, 2022, p. 155) is critical for communicators seeking to develop public trust; technical documents focusing on scientific objectivity can lack that embodiment, and "people understand the environment primarily through their physical bodies" (Le Rouge, 2022, p. 155) and ways they relate to it and other people. Reaching out to specific, influential community members can help with this embodied, trust-building process through relationship. Walton (2013), for example, and influence of trusted community members as she describes a farming project that was being introduced into a community by researchers: "Farmers who trusted the project told their friends, neighbors, and relatives" (p. 96) to facilitate the project within their communities. "Partnering with beneficiaries early in a project and ensuring that their input shapes the project focus can be an effective way to establish trust," Walton (2013, p. 99) states. These beneficiaries then developed a reciprocal trust with other trusted farmers who had partnered with researcher outsiders who were interacting with the community members so that these projects could succeed. Hartelius and Browning (2008) connect this process specifically to ethos:

> For example, the manager–rhetor is indeed subject to ethical—or ethos-based—concerns. His or her ethos is a function of an adaptive performance of character in any rhetorical moment. Ethos invites the audience members to grant credibility and trustworthiness to a speaker—or, in this case, a manager. (p. 29)

Hartelius and Browning indicate that this process is a reciprocal one: ethos, including the characteristic of trustworthiness, is a jointly constructed and negotiated process between the rhetor and audience. This negotiation process can be facilitated when participants can make connections between scientific language and evidence (that at first may seem unrelatable) and personal experiences and values. As Le Rouge (2022) points out, this type of "embodied understanding of the world" (p. 155) can be missing from technical communication. Importantly, all parties within a conflict need to develop and reciprocate trust.

Trustworthiness was not just a part of ethos that Edwards alone needed to negotiate with the community; the community would also need to demonstrate and negotiate its ethos with Edwards so that Edwards could trust the community to participate in the environmental protection efforts that both would ideally value, ones that would resonate with the community's strong, embodied experiences at Grey Cliffs.

ENVIRONMENTAL AND SOCIAL JUSTICE

The final part of the theoretical framework informing this case study is environmental and social justice theories; these manifest themselves throughout the discussion of narrative and ethos-building theories, since the topic of this case study highlights environmental sustainability and rural populations seeking voices in this conflict. Considering social justice alongside environmental justice provides space for a developing rhetoric of relationship ultimately required to protect Grey Cliffs environmentally. Originating from critical race theories, which "recognize the marriage between language and action" and "encourage critical thinking and engagement with language and social structures" (Edwards, 2018, p. 274), social justice theories likewise encourage attention to remedying power inequalities and facilitating diverse perspectives and action within various social systems and structures (Mooney et al., 2022, p. 134; Walton et al., 2019), including the use of counterstory (Martinez, 2020, p. 17), told through participants' "lived experiences" (Moore et al., 2021, p. 9). As Jones and Walton (2018) clarify regarding connections to technical communiciation specifically,

> **social justice research** in technical communication investigates how communication, broadly defined, can amplify the agency of oppressed people—those who are materially, socially, politically, and/or economically under-resourced. Key to this definition is a collaborative, respectful approach that moves past description and exploration of social justice issues to taking action to redress inequities. (p. 242, bold emphasis in original)

Social justice perspectives emphasize "intersectionality," which Walton et al. (2019) define as "an approach to understanding oppression that sees oppressive structures as intersecting, interlocking, and inseparable" (p. 12). This complex dynamic requires a collaborative, "coalitional approach" (Walton et al., 2019, p. 12) to begin addressing oppressive power structures; this process goes far beyond the individual level of experience and incorporates others' experiences and needs, humbly acknowledging the complexity of this social justice work and working together "not only to recognize oppression but also to reveal, reject, and replace it: To take action" (Walton et al., 2019, p. 50).

Within the framework of social justice, environmental justice focuses more specifically on integrating diverse perspectives and publics within decision making and participation in environmental issues, including "the necessity to focus on power as it relates to marginalized groups" (Sackey, 2018, p. 151), especially when risk is involved (Haas & Frost, 2017, p. 169; Williams & James, 2009). As Sackey clarifies, "focus on race and sustainability has always been consistent with organizers who use EJ as a frame. It is the meaningful involvement of people in the development, implementation, and regulation of environmental policy" (p. 150). Environmental justice also highlights the "emplaced, material, and environmental nature of social justice concerns" (Sackey, 2018, p. 139), and it can also include theories and discussions of ways to frame environmental risks in ways diverse audiences can understand and participate in (Haas & Frost, 2017; Stephens & Richards, 2020, p. 5), highlighting two-way communication between these audiences and policy makers rather than a one-way communication approach between governmental communicators, for example, and communities (Olman & DeVasto, 2020, pp. 17–18; Simmons, 2007). An environmental justice orientation includes "ascertaining and making apparent underrepresented and ignored groups who are stakeholders in contexts of risk, valuing the lived, embodied experiences of *all* stakeholders, and learning most from those for whom environmental equity is not an embodied reality" (Haas & Frost, 2017, p. 181). These approaches also address the nonhuman, such as environmental conditions of spaces/places and the animals and plants that inhabit them, "as injustices against any living species (not just humans and non-human animals) should impact the social" (Haas & Eble, 2018, p. 11). According to Butts and Jones (2021), environmental justice "suggests that people are an integral part of the environment and that to address environmental injustice requires that we also address its intersections with other forms of social inequity," such as the "income level" (Butts & Jones, 2021, p. 6) inequities that the Grey Cliffs community experienced. Complementing this perspective, scholars (Lindeman, 2013; Simmons, 2007; Tillery, 2017, 2019) suggest a need for and growing presence of local citizen involvement in environmental and public policy decision making, although Lindeman (2013) and Tillery (2019) voice some cautions that public persuasiveness can counteract scientific expertise in potentially damaging ways.

The concept of intersectionality provides room to consider marginalization for communities such as Grey Cliffs based on positionality, power dynamics, the need for environmental equity, and other forms of nondominance. For example, the rural Grey Cliffs community members, while white, experienced varying degrees of economic marginalization. The public schools in the county where Grey Cliffs is located have been designated as Community Eligible Provision schools, and students attending receive free meals based on overall household income levels. From year to year, a couple of schools have moved in and out of this category, but most students qualify for food assistance. In addition, many households in the area participate in the Affordable Connectivity Program to receive internet access at reduced cost based on household income in this rural area. Many residents have not attended college, and, while they are experts in their workplaces and trades, their lack of experience interacting with government organizations such as the Corps fostered reduced confidence and confusion about how to address the Grey Cliffs conflict, which was based on government-established regulations and property. In addition, rural populations' voices may not always be acknowledged by industries and organizations (Carlson & Caretta,

2021, p. 41), leading to further silencing, when these community stories need to be heard (Haas & Eble, 2018; Walton et al., 2019). This community, while not orally expressing marginalization in these terms, seemed to intuitively understand that their values, experiences, and stories differed from the Corps in ways that positioned them at a disadvantage compared to the all-powerful Corps as a government organization.

These intersectional complexities surrounding such a rural community serve as motivations to connect technical and professional communication, organizational communication, and environmental communication to the fields of social and environmental justice, to change the conceptualization of technical and professional communication as a "means to an end" and instead connect these fields further to "political and social effects" (Jones & Walton, 2018, p. 248). These effects illustrate how technical and professional communication and communication fields apply to specific, local contexts and conflicts (Agboka, 2013; Sackey, 2018, p. 139; Simmons, 2007; Tillery, 2019) as well as "the everyday in situated locales" (Sackey, 2018, p. 139). Importantly, as this case study demonstrates, "In terms of EJ, groups on the ground always have agency because their movements are driven by their own goals and experiences. The only limits placed on their agency stems [sic] from unequal power relations between themselves and either corporate entities and/or government agencies" (Sackey, 2018, p. 153). However, historically in the United States, environmental discussions have foregrounded dominant narratives and marginalized others: "This privileging legitimizes certain groups while delegitimizing those groups pushed outside the boundaries of (or marginalized by) the discourse's dominant narrative" (Mangum, 2021, p. 57). Having discussions about environmental concerns and policies that are truly just means that "all affected by the decision [have] the ability to actively participate in the decision-making process" (Simmons, 2007, p. 3). Simmons (2007) recommends that, for successful policy debates involving the public, the public should be encouraged to participate early on in the process, and officials should view the public's knowledge as valuable (p. 4). In addition, officials should take the time to learn about the public's values related to specific environmental concerns and even take their emotions into consideration when planning communication and public involvement (Grabill & Simmons, 1998). The multiple narratives gathered for this study illustrate these unequal power relations related to environmental issues yet also indicate the potential for agency when diverse narratives and stories are respected by those in power. Within these narratives, the connection to Grey Cliffs as a place is critical to motivating the community members to accomplish action. As Sackey (2018) writes, "*Space and place is always local and always material in EJ movements. This focus is always driven by the needs, demands and goals of citizen groups working to solve problems within their communities*" (p. 152, emphasis in original). In this context, stories and counterstories function as a way to not only "transmit local experiential knowledge to inform the public" (Baniya & Chen, 2021, p. 76) but also "provide ways to navigate and survive a crisis" (Baniya & Chen, 2021, p. 76).

Within this case study, all parties played an important role in co-constructing ethos and this environmental preservation narrative; in order for it to work, it couldn't be a narrative promoted by the Corps alone. The narrative couldn't be one promoted by the community alone, either, since narratives about Grey Cliffs' virtues certainly weren't enough to address the problems the Corps needed to take action on. Instead, the Corps and community needed to construct a common narrative and rhetorical framework for action. In addition, "a shared narrative provides the fundamental social cohesion within any organizational environment. It is a rhetorical form of identification and collective purpose" (Hartelius & Browning, 2008, p. 31). In essence, this shared narrative would be the "glue" that would socially bind these governmental (organizational) and community social actors together as they worked together to keep Grey Cliffs open through meaningful relationships based on trust and aligned values.

Through this observational case study, I view and analyze all of the data collected, including the transcribed interviews, through these theoretical and rhetorical lenses to emphasize the need for a rhetoric of relationship during this environmental conflict. Although viewed through my own limited interpretive lens, these varied data present community and governmental narratives that reveal much about the process of negotiating agency within a very tense rhetorical context where much was at stake from all perspectives. Chapters 3, 4, 5, 6, and 7 present the results and discussion of this ethnographic, observational case study, as analyzed through these theoretical lenses. I refer to these concepts throughout the next chapters, although some apply to particular communicative events during the conflict more than others. Chapter 3 applies this framework specifically to analyzing community narratives to identify community stories and values that characterized its relationships with the Corps, other community members, and the Grey Cliffs environment during this conflict.

Laid grandplan for communication

3
COMMUNITY NARRATIVES AND ETHOS
Agency and Values

"Because these people were all raised there, they were kids, they got baptized there; they've been baptized there, uh, you know, church used it for baptisms, you know."

—Paul, community member
and General Store owner

"We may have picked up a few things, but nothing like. And all this about needles and stuff—I've never seen or never picked no needles up."

—Tom, community member and
town hall meeting participant

This chapter focuses on identifying community values that proved essential to understanding more about the value alignment process between the Corps and community, in spite of the communication difficulties they experienced. Through this observational case study and analyzing community members' reflective self-narratives, gathered through participant interviews, I present what some of these values and embodied experiences are; these "anecdotes about the virtues of a way of life" (Patterson & Lee, 1997, p. 29) are clues to some ways these values could potentially align with Corps values eventually. These values are culture specific; while aspects of these might apply to other audiences in general, they can also apply in audience-specific ways, both in relation to the audience as a whole and as individual actors. In this case, community members' values combined to form a group ethos: while individual community members revealed these values through their own unique stories, these values combined through members' relationships to each other to demonstrate a negotiated community ethos that bound them together against threatening outside influences, such as the Corps.

COMMUNITY VALUES

I grouped these community values into the following categories:

- Value of religion
- Value of tradition
- Values of recreation (and its convenience) and family time
- Value of skepticism of government authority in its representation of fact
- Value of social unity, indicated by respect and the need for all voices to be heard

All of these values necessarily relate to trust, since, if communicators and their audiences demonstrate that they share the same values and vice versa, they can negotiate and build trust with one another, fostering not only effective communication but also jointly negotiated social action. See Table 3.1 for codes I used to identify these values, based on the grounded theory analysis.

Value of Religion

Located in the Bible Belt of the United States, the Grey Cliffs area is located in a politically conservative, religiously active community. The primary religion practiced by community members is Southern Baptist, indicated by the large numbers of Southern Baptist churches in the area. One Southern Baptist church is located just a mile away from Grey Cliffs, and members use the lake access for baptisms within their congregation. One of the most valued religious practices of these churches, baptism is practiced through immersion in water. Seeking to imitate Christ's example in the book of Matthew (3:13–17), members of the Southern Baptist church believe that immersion is really the only way to follow Christ in baptism. Because some of the churches in this area are so small, many do not house their own baptisteries. As a result, some communities such as Grey Cliffs will use locally available lakes and other bodies of water, such as pools, to use for these baptism services.

Denise is a member of one of these churches in the area. As Tom, her husband, was narrating the benefits of Grey Cliffs, Denise mentioned, "I got baptized in there." Tom confirmed, "Yeah, and I said, you know, they have a lot of baptisms; [the local Southern Baptist] Church has a lot of baptisms down there, you know, all the time; they use that." To Tom and Denise, Grey Cliffs had a strong religious significance and importance to the community. This use of the area for religious purposes was part of the community's justification for keeping Grey Cliffs open. While none of the interviewees mentioned the concept of freedom of religious

Table 3.1. Coding Scheme for Community Values

Code (value)	Explanation	Example quotations
Religion		
	Statements connecting Grey Cliffs to religious activities, such as baptism.	"I got baptized in there." "Yeah, and I said, you know, they have a lot of baptisms; [the local Southern Baptist] Church has a lot of baptisms down there, you know, all the time; they use that." "Because these people were all raised there, they were kids, they got baptized there; they've been baptized there, uh, you know, church used it for baptisms, you know."
Tradition		
	References to the cultural history of the area regarding accepted practices and ways activities are normally carried out.	"Consider[ing] their, um, their history, um, in the, you know, people grew up in this area, they grew up using that area, they've taken their, they've grown their kids up in that area, um, and so it's very important to them to keep it open." "They only worry about the dollar." "And me, I worry about people's feelings, people's you know, what they're eating, how, quality of the food it is." "[Grey Cliffs] belongs to the community . . . these people were all raised there." "My forefathers and my family before the dams were brought in, they used to fish there a lot. Yeah, and they used to camp underneath the falls."
Recreation, convenience, and family time		
	References to recreational activities at Grey Cliffs, the convenience of the location or inconveniences of having it closed, spending time with family there.	"My dad used to bring me and my sisters here to learn how to swim. And we always went down there to swim; I always used to go there and put a boat in, fish, which we'd fish a lot of the night and everything, you know, and then when it was too hot to fish or something we'd, a bunch of us get together and go down there and water ski and swim and ah just hang out." "I was raised, probably, what, a mile and a half from [Grey Cliffs]. And I grew up there all my life; actually we used to go down there and swim all the time and, uh, you know lot of the neighbors, several of them, used to go down there and take baths in the winter time, summer time, um, people go down there and go fishing." "I used to walk down through there, where you're talking about, go down through to [Grey Cliffs], and pick blackberries." "It's convenient to be able to go there, take the kids there, swimming, so on. Fishing."

continued on next page

Table 3.1—*continued*

Code (value)	Explanation	Example quotations
Skepticism of government authority		
	Statements about the Corps not doing something it said it would do. Statements about government officials' statements about nefarious activities taking place at Grey Cliffs as "lies."	"Yeah, the focus was on closing it, and uh, the focus at first was to actually close it for two years, but the opinion of all those who live around here, the um, the community, was that if it closed down at all, it would never open again, and I kind of agree with that, you know, that that's probably what would happen." "I've never, I've never see, I told [Edwards], I said, I've never seen nobody that I've been afraid of." "I think that, uh, a lot of us have learned that uh if we leave it to the authorities and we leave things unchecked, um, they will, you know, they'll just do whatever they feel is easiest for them." "I called him [Edwards] out, you may know, you may have remembered, you know; I called him out on a few things; and I told him I just thought it was bare faced lies." "We may have picked up a few things, but nothing like. And all this about needles and stuff—I've never seen or never picked no needles up."
Social unity		
Respect	Statements including the word "respect" or references to being treated well by government officials, as if community members' concerns were valid. Other statements relating to respect found in community members' stories. Statements referring to respecting Edwards.	"But she [Denise] called [Tammy Phillips, a government representative], you know, and I said, 'Honey, she will not call you back.' Within 15 minutes she sure did." "This man [a business owner] disrespected a 78-year-old man that needed help on a 90-degree weather; we don't do this in [name of city]; we try to help elderly people. . . ." "He should have not disrespected an elderly person that needed help; he should have helped him. And talked to him the way he did; we don't do that here." Paul was "training [his daughter], you know, to be self-respecting to people, as well." "Yeah, I have a personal interest in making him [Edwards] happy because he's, um, the authority in the Corps."
Need for all voices to be heard	Making sure "old timers" or those without internet access were included in community conversations. Listening to others' points of view.	"Yeah, I mean, cause [a] lot of older community members, uh, you know, a lot of people are still not on social media. Uh, and they get their word of mouth by the old stores, you know like they used to, you know, their communication's still going to have a cup of coffee with somebody and talking to Mr. Joe, and finding out what the week's scoop is." "I kept my mouth shut the first couple of meetings, just to hear everybody out. And then from there, I felt I could, um, voice and form my opinion based on the feelings and the thoughts of others, what they really wanted to see."

expression that Grey Cliffs facilitated, the implication was that if the area were closed, these community members could no longer conveniently practice parts of their religious services and beliefs. Especially because the Corps was a government organization, the community's implication that a government action might inhibit religious freedom sent a strong message to Edwards and others involved during this negotiation process, such as a local business that was considering donating funds, equipment, and material to help clean up the area. Tom's mentioning that Grey Cliffs was used for this religious purpose in effect introduced Edwards and the local business to this important community value.

Value of Tradition

The Grey Cliffs community was also one composed of families who had lived in the area for generations. As a result, interviewees talked about many different traditions that they valued. One of these traditions was simply visiting the area. Lee narrated:

> Uh, like I said, consider[ing] their, um, their history, um, in the, you know, people grew up in this area, they grew up using that area, they've taken their, they've grown their kids up in that area, um, and so it's very important to them to keep it open.

Although not from this area originally, Lee was one of two families whose land bordered the Corps land that then bordered Grey Cliffs. He was one of the people who often had his property trespassed upon by people who were off-roading at Grey Cliffs, who would leave the Corps property and access adjacent farms. Interestingly, even though Lee would have greatly benefitted from Grey Cliffs' closure, since trespassing on his property would then be eliminated, he sympathized with the community members who wanted to keep it open. While he enjoyed using Grey Cliffs himself, he could have valued limiting access to his property more by supporting closure, and because of his land ownership, he could have presented a credible case. Understanding the community's emphasis on its tradition of using the area, though, partly because he was a longtime (not lifelong) resident of the area and had observed the community's dedication to the area firsthand, Lee agreed to support this tradition of keeping the area open and emphasized the importance of access as he communicated with the other parties involved in this conflict, including Edwards. Lee emphasized this community value through his role as an executive committee board member, a role he assumed when the community developed a board to begin a more formal process of negotiating with the Corps.

Another community member involved in this conflict was local business owner Paul. During his interview, he referred to traditional values of caring for others that he attempted to convey through a general store he had recently built about 1.5 miles from Grey Cliffs. "See I, I, everything I do is for a meaning," Paul stated, as he told me about modeling his general store's construction after one that existed nearby when he was a boy. At this store, Paul remembered, the owner would often feed him for free and keep an eye on him while he ran the store, just to show his support of a young boy. Paul valued his experiences at this other store as a child and wanted to create a similar environment for community members and others visiting Grey Cliffs. This general store became the location of the town hall meetings about preserving Grey Cliffs. In contrast to his business philosophy that was more personal, he said, "They only worry about the dollar," referring to chain stores, such as Dollar General and chain restaurants nearby. "And me, I worry about people's feelings, people's you know, what they're eating, how, quality of the food it is." Paul also mentioned his willingness to donate water to people just coming into the store after hiking in the area or participating in Grey Cliffs cleanup efforts. When I asked him if the store was doing well financially so that he could easily donate these things, he said, "It ain't a lot of money [that he gets from running the store]. 'Course I need to keep the lights on, but other than that, I don't care." Trying not to "forget where he came from," Paul's desire was to promote values shared by others in this primarily working-class area, values such as the ones he mentioned that are associated with a simple life. Of course, when the possibility of closing Grey Cliffs was first mentioned, Paul was probably nervous that his million-dollar investment could be jeopardized by fewer people visiting the area. However, when I questioned him about this, he said,

> Well, obviously, [Grey Cliffs] belongs to the community, uh, you know, and I've been out, even though I was here as a kid and whatever, my family's been here, I've been out of the country for a lot of years. And, you know, things change, and, um, I've been gone for like thirty plus years, um, so I wanted, this belonged to the community; you could tell how passionate they were, so I wasn't about to, uh, to go in and, and uh, say anything [about the impact on the store], you know, because these people were all raised there, they were kids, they got baptized there; they've been baptized there, uh, you know, church used it for baptisms, you know, and I wanted all of that out first. It wasn't about me or what I was planning on doing.

Paul realized the values this community promoted, such as the religious ones, and he knew those took priority over others, such as an individual making money from a newly constructed business. Notably, Paul knew that the community valued trust and that, as a returning resident, he would need to gain trust with the community as a local business owner as well. Not focusing on money as a business owner but instead providing a supportive location for the town hall meetings and other items, such as food and beverages for the meetings, was a way to begin negotiating trust with the community. He knew that focusing on profit alone would reveal self-serving motives that the community could not trust. Because Paul was from this area and his family continued to live in the area after he had left, he was familiar with these traditional community values, and they governed what he said and didn't say during the meetings.

Values of Recreation, Family Time, and Convenience

For this community, Grey Cliffs was a place to recreate locally, where families could enjoy spending time together. Because some families have lived in this area for generations, they had longstanding family memories of sharing special times in the area. Paul recollects:

> Well, actually, it was um, my forefathers and my family before the dams were brought in, they used to fish there a lot. Yeah, and they used to camp underneath the falls, when it wasn't a state park. Yeah, so it's actually, the family's been, uh, fishing this area for a long time. When it was a river. Uh, 'cause like I said, I knew the place since I was a kid; my family was there; I knew the place quite well 'cause the [family surname] used to fish all the lands here.

Paul, especially, would understand the problematic relationship of the Corps and the community, since his family was in the area "before the dams were brought in," before Corps involvement, when individual landowners truly did own the area. Paul's narrative indicates that even before the Corps bought the land from the local landowners and created the lake, the community had a sense of ownership of the area because this area constituted families' homesteads. Several other community members reflected that sense of ownership as it related to values of recreation and family time. Tom also narrated stories of times his family used the area for recreation, and his wife Denise did as well. Denise and a friend, Felicia, also mentioned picking blackberries in the area during the summer. Felicia said, "I used to walk down through there, where you're talking about, go down through to [Grey Cliffs], and pick blackberries."

In addition, the community members I interviewed valued the convenience of Grey Cliffs' location. For example, Lee mentioned, "it's convenient to be able to go there, take the kids there, swimming, so on. Fishing." And Denise, Felicia, and Paul all mentioned their families living in the area and visiting Grey Cliffs often. Clearly, these community members valued the recreational opportunities the area offered, the family time they experienced there, and the convenience of Grey Cliffs to where they lived.

Distrust or Skepticism of Government Authority

According to Thomas et al. (2009), trust can be defined as "part of a relationship between two people and involves the voluntary acceptance by the trustor of risk based on the actions of the other party" (p. 290). These authors assert that "despite the importance of this topic, the current literature provides little guidance for managers on how to use communication as a means to increasing levels of trust. This study shows that to increase trust among coworkers and supervisors, it is important that information be timely, accurate, and useful" (p. 305). In this case, the Grey Cliffs community certainly wasn't willing to take any risks with Edwards because the information Edwards was presenting didn't seem to fall in any of these categories for the community. As Lee said,

> Yeah, the focus was on closing it, and uh, the focus at first was to actually close it for two years, but the opinion of all those who live around here, the um, the community, was that if it closed down at all, it would never open again, and I kind of agree with that, you know, that that's probably what would happen.

Community resistance originated from the belief that what the Corps said it was going to do would not actually happen. Part of this distrust also related to continued suspicions developed over time as the Corps had appeared to be an organization that valued actions taken for its own benefit rather than the benefit of community members and landowners who had no choice but to sell their land to the Corps for management of the lake. This distrust harmed Edwards's ethos in the eyes of the community, especially his credibility, since he was closely related to a government organization that had historically been viewed as betraying the community.

In addition, the Corps' focus on the violence taking place at Grey Cliffs seemed blown out of proportion based on previous community experience with the area, and this contrast in perception created more distrust in the community. Interestingly, Tom, Denise, and Felicia, who all three grew up close to Grey Cliffs, told stories about violence there in the area's past that made the current crime seem less significant. Denise stated:

> And it's been, it used to be a lot rough, rough people. But all them people's dead and gone. You know. But it was mostly like brothers stabbing

brothers or brothers cutting brothers or you know, families fighting, it wasn't just like you go down there, and us attack you, you know; it was families; it was all families. Fighting each other.

Complementing Denise's memories, Tom talked about the killings being worse when they were younger, in contrast to what Edwards was talking about now. To these community members, the violence had been so bad before, even though it was family violence and not random violence, that whatever violence was happening at Grey Cliffs currently wasn't nearly as severe and therefore did not warrant the drastic response that Edwards was proposing. Community memories, voiced through counterstories, differed from Corps memories, and these differences impacted various perspectives on how to address these concerns, all tied in ultimately to who owned and accessed this area, who could be trusted to make judgment calls about decisions related to what would happen to Grey Cliffs and who could access it.

Although not originally from this area to share the common history of the community's relationship with the Corps, Lee also communicated his broader perspective on the government's role in the Grey Cliffs conflict that at first hinted at distrust but then presented a more positive view of the Corps:

> Umm, I think that a lot of us have learned that if we leave it to the authorities and we leave things unchecked, um, they will, you know, they'll just do whatever they feel is easiest for them, which I don't blame them, but if we as a community will speak up, um, work with the authorities, then, I believe that they will, you know, take heed to that, you know, they, I think they'll work with us. So I think, I think it's important to continue to communicate, you know, with a civil tongue, communicate in a way that's, that's, um, considering their job and their responsibilities, um, and uh, communicate what our desire is for our community, so. I think it's important.

Lee's words "communicate, you know, with a civil tongue, . . . considering their job and their responsibilities" suggest another value that was very important to this community: respect. As I learned through analyzing the interview transcripts, this complicated value manifested itself in various dynamics, including government officials' treatment of community members, the community members' treatment of government officials, and other seemingly tangential stories that highlighted the importance of this value to the community even more. Respect was such an important value to this community that it may very well have been the most important one and could further explain the strong distrust the community members felt toward Edwards and their hostility toward him, as they sensed a possible lack of respect for community feelings and ideas at first. This value of respect ultimately connected to a broader value of unity, which also entailed allowing all community voices to be heard.

Value of Social Unity, Indicated by Respect and the Need for All Voices to be Heard

In this close-knit community, composed of individuals who have lived in the area for generations and shared a common history, particularly as it related to Grey Cliffs, community members revealed through their self-narratives that unity was also valued. While the community did not use the term "unity" to indicate this value, they did discuss related themes that all contribute to social unity: respect and the need for all voices to be heard.

Respect

Essentially, when the Corps bought out farms and land from local landowners to create the lake, the Corps' actions could have been interpreted as disrespect. As Edwards had indicated, "antigovernment sentiment" formed in this politically conservative area when disagreements erupted not only about the Corps land buyouts but also about the compensation landowners received. Many felt this process had decimated their families' livelihood and plans to leave the land as inheritance for future generations, and these families left the area (Williams et al., 2016). These actions had fragmented this once-unified community, and this history had not been forgotten. Therefore, when Edwards approached the community with the proposal to close the area, even temporarily, this action could have been seen as disrespecting the community and threatening the community's unity even further.

Although Edwards didn't fully realize this and did not intentionally want to indicate disrespect for this community and its interests, his previous lack of communication with the community about Grey Cliffs and its problems implied disrespect. Even though his first conversations with selected community members about closure indicated support for temporarily closing the area, the community as a whole did not support this option and had no idea this possibility even existed. Edwards's suggestion in itself, then, introduced more disunity, which became strongly evident at the first town hall meeting.

In contrast to these early interactions with Edwards as a government representative, though, was the response of a local congresswoman, Tammy Phillips, to Denise when Denise called her to express concern about the conflict and to ask what Phillips might be able to do to help keep the Corps from closing the area. The surprise Tom and Denise experienced when Phillips almost immediately responded to Denise's phone message was evident during my interview with both Tom and Denise:

TOM: "But she [Denise] called [Tammy Phillips], you know, and I said, 'Honey, she will not call you back.' Within 15 minutes she sure did."

DENISE: "She did, too!"

TOM: "She called back personally."

DENISE: "Personally, her own self."

TOM: "Her self, not no representative."

DENISE: "She didn't tell somebody else; she called me."

TOM: "Talked for an hour."

KP (INTERVIEWER): "Um, and so you talked to her about that—"

DENISE: "Yes, yes."

KP (INTERVIEWER): "What, how did she respond to that?"

DENISE: "Uh, she told me, she said that she had heard a little about it, she said but, you know, that's ridiculous. She said uh—"

TOM: "Yeah, she didn't think—"

DENISE: "She told me that [Bob Wheeler] and [another congressional representative] was the one I need to talk to. She said, but she said, I will put in a good word for you, and she said, I will send a representative, she said."

TOM: "She did, too."

DENISE: "If I can't make it myself, she said, the next meeting you have, you let me know, and she said, I will personally send a representative to represent."

KP (INTERVIEWER): "That's awesome; yeah, I remember seeing [Tammy Phillips's representative] at the first meeting."

DENISE: "Mmm Hmmm."

TOM: "Yeah, [Tammy Phillips's representative], yeah."

DENISE: "Yep, sure did."

This personal interaction Denise experienced was obviously important to both Denise and Tom, as they both excitedly narrated. Skeptical that Representative Phillips would respond or even care about this situation, Denise and Tom went on and on about how Phillips had not only called Denise back but had also sent a personal representative to the meetings. This interaction indicated respect for Denise and her concerns; this support from a government representative also suggested a type of support for this problem that had disrupted the community, especially when

Phillips said that she thought this situation was "ridiculous," supporting the overall concerns of the community.

In addition to this support from a government representative, another illustration of community members' valuing respect and unity was revealed in Tom's seemingly unrelated story included in his narrative about helping an elderly man who was stranded in a parking lot nearby due to a broken-down car. Tom, whose auto shop was located close by, attempted to help the man, when another man working with a nearby business began harassing them, insisting that Tom and the elderly man remove themselves and the disabled vehicle off of his property immediately. The insensitive treatment offended Tom so much that he took pictures of the offending man, which included the physical location of the business itself, and he posted the story on social media. Tom recounts:

> So I put on social media, now this is the power it's got on social media, that this man disrespected a 78-year-old man that needed help on a 90-degree weather; we don't do this in [name of city]; we try to help elderly people. This is not the way, what [name of city's] about. That this guy is evidently not from here; and he wasn't; he was from New Jersey. And I said, you know, anybody that does business with him, is, I thought he owned the cash company, like uh, what was the name of that place? Cash, it wasn't Cash Express, whatever it was, I said, is as sorry as he is. I mean, I, we wrote up a beef on it. I really don't care; he should have not disrespected an elderly person that needed help; he should have helped him. And talked to him the way he did; we don't do that here.

Later, Tom received a call from a woman who owned the flower shop next door to the check-cashing business; she was also the district manager of this check-cashing firm. She said that they rented their office space from this man from New Jersey and that the social media post, including the pictures, had ruined both of their businesses. People did not want to be anywhere near a business that would disrespect the elderly in such a way.

When Tom learned how his post had brought harm to an innocent business owner next door, he apologized to the woman on the phone and then posted an apology to social media.

> I said [to the flower shop owner], no, I owe you an apology, you know, 'cause I didn't mean to hurt your business. She said just as soon as our lease is up, we're out of there. I mean I don't know if they're there no more or if they moved. But I sure didn't know it would hurt, I mean, that it would hurt their business, you know, that social media would have done that, but it did. And I apologized to her, and I said, look, you know, I'll. And I did, I put another comment, you know, hey, I'm not here to hurt these people; this man is what I want to get across.

Although Tom had acknowledged that this scenario had nothing to do with the Corps situation we were discussing, it did connect because Tom was emphasizing how much he had learned about the power of social media: based on this experience, he was now using social media for positive purposes to increase people's awareness of the changes going on at Grey Cliffs. But also, importantly, this story emphasized the importance of the value of respect, again. And consistent respect for others was a value that unified this community. While this story elevated Tom as he respected this elderly gentleman who needed his help, the story also elaborated on a lesson Tom learned from this experience: social media has great power and should be used cautiously and appropriately.

For Tom, this counterstory addressed the "dominant cultural narrative" of growing disrespect within the culture he was familiar with, whether that was Corps involvement or the disrespect of individuals such as this elderly gentleman from an outsider. This story, as well as the one Tom and Denise told about Congresswoman Tammy Phillips, deserved repeating because they spoke against a narrative of disrespect that could fracture unity within this community. Repeating these stories gave Tom and Denise power as they challenged the dominant cultural narratives around them.

As Heath (1983) further explains,

> Common experience in events similar to those of the story becomes an expression of social unity, a commitment to maintenance of the norms of the church and of the roles within the mill community's life. In telling a story, an individual shows that he belongs to the group: he knows about either himself or the subject of the story, and he understands the norms which were broken by the story's central character. (pp. 150–151)

While speaking about a specific working-class community and oral traditions practiced by it, Heath's description of stories within a community applies to the Grey Cliffs community as well; these stories serve to maintain norms and values, and they reproduce "templates" of roles community members play as they help others and work together. In addition, in Tom's case, he broke a social norm through his caustic social media post against a business, and he states he learned from that. As discussed in Chapter 7, Tom now uses social media to unite the Grey Cliffs community in its relationship to the new activity sanctions and regulations there.

In addition to these discussions of respect revealed in these community members' self-narratives, two other community members referred to this concept. Paul indicated that he was "training [his daughter], you know, to be self-respecting to people, as well." His daughter, who was about five years old at the time, would one day inherit the family business, and Paul wanted to train her in this value early. And Lee, while not mentioning respect directly, indicated that he wanted to retain a good relationship with the Corps so that he could continue accessing his farmland, which he needed to cross Corps land in order to access.

Allowing All Voices to Be Heard

Another type of unity in this community was an effort to ensure all voices could be heard, a value referred to generally by Paul as he talked about the value of his business as a gathering place for the town hall meetings and the community in general. At the first meeting, so many community members attended, most objecting to closure. That meeting was a time for all community members to voice their views, and Paul believed that the loud voices of the community surprised Edwards. Several interviewees had mentioned they thought Edwards was expecting a very low turnout at the meeting, and then the Corps could proceed with closing the area if that seemed to be an appropriate choice. However, the community countered this possibility with numerous voices and opinions about what should or could be done. This community valued the opportunity for all to speak, which in this case extended to selecting a physical location, such as Paul's general store, where that activity could take place.

Paul took pride in practicing what he considered traditional values, such as the ones described here about open communication; he saw his general store as an embodiment of some of those values, such as when he spoke of providing a place for families to gather with their kids, hosting fundraising events, and serving as a location where "old school" communication took place, such as "front porch" conversations and posting fliers advertising upcoming meetings and other community events. He also mentioned several times that not everyone in this community had internet access or was connected to social media. These community members, some of them "old timers," were important as well, and Paul wanted to make sure they were included in community conversations and involved in the issues impacting the area. He stated:

> Yeah, I mean, 'cause [a] lot of older community members, uh, you know, a lot of people are still not on social media. Uh, and they get their word of mouth by the old stores, you know like they used to, you know, their communication's still going to have a cup of coffee with somebody and talking to Mr. Joe, and finding out what the week's scoop is.

While Paul kept an active Facebook page that advertised events at the store and encouraged visitors to the area to visit both the store and restaurant, his ethos was one of welcoming those who could not access this information via social media. His desire was to create a place where all voices were welcomed, and he practiced this philosophy by opening up his store to host the town hall meetings. Even outsiders were welcome in this environment; during the first meeting, one man spoke up, offering his opinion about the situation, and longtime residents accusingly questioned the unfamiliar man if he was from the area. Acknowledging that he was from California and had actually been living at Grey Cliffs, he took the risk of being shunned by the community. But Paul, who also owned a campground adjacent to the store, provided him with a place to stay when living at Grey Cliffs was no longer feasible for this man. Some might say Paul's efforts all revolved around the value of unity, facilitated by various forms of communication.

Lee focused on unity by practicing a different communication strategy: not immediately stating his opinion about what should happen at Grey Cliffs. Lee had lived next to Grey Cliffs for 20 years and understood the community sentiment about the area, even though he was not native to it. While respecting the voices of the insiders, Lee calculated how he as somewhat of an outsider could help the community and make a difference:

> I kept my mouth shut the first couple of meetings, just to hear everybody out. And then from there, I felt I could, um, voice and form my opinion based on the feelings and the thoughts of others, what they really wanted to see, like I said, I'm willing to go either way with it, but there is a lot of people who, it really mattered to them keeping it open, and so my opinion leans more towards keeping it open for the sake of others, I guess, so.

Integrating others' opinions and allowing others' ideas to form his own views allowed Lee to participate in unifying a community that found itself in a very vulnerable spot in relation to a powerful government organization, the Corps.

The values of respect and allowing all voices to be heard connect closely to the overall value of unity. For example, Denise wanted a government representative to hear her voice and respect her experience, her Grey Cliffs narrative. Tom's seemingly unrelated story about helping the elderly man in need demonstrated respect for this man but also for his voice, his words expressing a need for help. Tom also grew to realize his need to respect a business's reputation via social media; while all voices do need to be heard, Tom realized that his posted story degrading the businessman's actions was hurting the business of someone else who was not involved in disrespecting the elderly man, and Tom chose to remove the story and replace it with something more positive. While

Tom valued his choice to post what he wanted at first, he also valued his choice to replace the original post with a different narrative for the sake of unity. Finally, Paul valued the opportunity for all community members to speak, even the outsiders. He acknowledged that not everyone has access to the internet and social media and made efforts to include them. These reciprocal community practices of demonstrating values through narrative and concrete actions solidify the strong foundation that initially found itself misaligned with Corps values.

COMMUNITY NARRATIVES AND ETHOS: GREY CLIFFS OWNERSHIP

These community values (religion, tradition, recreation and family time, skepticism of government authority, and value of social unity), conveyed through narratives and stories, demonstrate significant social capital for this community as part of its ethos; these common values are strengthened by a common physical heritage and connection to Grey Cliffs, including similar embodied experiences at this location expressed using common affective values. As a somewhat economically disadvantaged community that had experienced past loss of livelihood through Corps land takeovers, community members had built up much social capital that connected them all through a strong, common ethos, an ethos strengthened by its unique values and evidenced by generations of experiences and love for the geographic space that is Grey Cliffs. This community's social capital did not have anything in common with Edwards's most obvious cultural capital (education, military training, and experience), though, and this lack of connection contributed even more to the lack of forward movement in this conflict. The community's social capital did not seem to lend any agency toward negotiation efforts, especially since the community was implicated to some degree in the Grey Cliffs degradation process. In contrast to the community's ethos, conveyed through its values, Edwards's ethos, based on the capital he brought to this experience, presented itself as a dominating one through the rules and regulations Edwards had the power to enforce. As a result, the community's social capital did not appear to have any bearing on the seeming inevitability of Grey Cliffs' closure, although the community's antenarrative, which conflicted with the dominant Corps narrative, certainly could contain negotiating potential.

In addition, as Bourdieu (2007) has observed, this type of initial, impersonal relationship between Edwards and the community made any type of "transaction" (p. 173) very difficult. This type of "social distance"

(p. 173) also could require some type of formal agreement among the parties involved in this conflict in order to ensure the area could stay open rather than rely on word-of-mouth agreements. Such a formal agreement, in addition to the rules and regulations, could build up even more social distance between Edwards and the community, who might interpret such a formal, written agreement as verification that these community members were, indeed, "the guilty ones."

Chapter 4 presents Edwards's efforts to persuade the community that some action needed to be taken to protect Grey Cliffs, even action as severe as closure. Edwards's rhetorical focus at this point remained on credibility, highlighting his experience and expertise as a Corps of Engineers resource manager. In this authoritative role, sanctioned by the government overall and the Corps as an organization, he spoke from the perspective of enforcing rules and regulations, and these contributed to his constructing an ethos as a regulator.

KEY RECOMMENDATIONS FOR TECHNICAL, PROFESSIONAL, AND ORGANIZATIONAL COMMUNICATION AUDIENCES

- Learn the values of community members involved in decisions, based on organizational changes.
- Listen to community members' narratives and stories to help identify those values in ways that incorporate "being there" (Rai & Druschke, 2018, p. 4) for those audiences.
- Find ways to bridge technical, scientific communication and the affective values displayed by community members, including implementing accessible, inclusive language.
- Consider ways community ethos might be represented in technical and scientific documentation.
- Reflect on underrepresented populations impacted by technical, professional, and organizational communication and decision making; how can the poly-vocal characteristics of these voices be preserved within conflicts, discussions, and decisions?
- Identify community counterstories as sources of value nuances that can contribute to connecting with audiences in more inclusive ways.

KEY RECOMMENDATIONS FOR ENVIRONMENTAL SCIENCE AND PUBLIC POLICY COMMUNICATION AUDIENCES

In addition to the recommendations above, these specifically apply to environmental science and public policy communication audiences.

- As subject matter experts, determine ways environmental science communication and public policy may be focusing too much on technical,

scientific, and abstract language. Determine ways this language could be revised so that more nonexpert and community audiences can understand it.
- Identify ways environmental science communication and public policy might be more "localized" to include community members' perspectives and values related to suggested changes and improvements.
- Provide opportunities for impacted communities to not only voice concerns in a conflict but participate in the change process, such as by listening for (not just to) counterstories (Mangum, 2021).
- Consider ways community values not only might have exacerbated an environmental problem but could contribute to specific solutions.

4

MOTIVATING THE COMPLIANT INDIVIDUAL
A Corps Resource Manager's Rhetoric of Regulation

> *"So by . . . designating sites and making sure we're checking those sites, you're going to have a more responsible, compliant, um, person who understands that we know who used that site, we're monitoring who's using that site."*
>
> —David Edwards, Corps of Engineers resource manager

> *"You do know there's cameras down there, don't you?"*
>
> —Tom, community member and town hall meeting participant

David Edwards, in his role as the U.S. Army Corps of Engineers resource manager, supervised the Grey Cliffs lake-access point, and maintaining the area was his responsibility. From Edwards's perspective, serious legal issues were involved with maintaining Grey Cliffs. Using the discourse of rules and regulations, Edwards was required to enforce Corps regulations at Grey Cliffs; those were nonnegotiable. Some of these regulations governed land use based on sustainability issues. Of 41 lake-access points, Edwards stated that "[Grey Cliffs] is the one that had the most . . . negligent, nefarious activity"; this activity included crime, drug use, beer bottles and trash littered across the area, and off-road vehicle use, to name a few of the more serious activities. According to a county dispatch report that Edwards cited, over a 2-year period, 90 pages contained documented calls received for problematic activities occurring in the area. All of this activity violated the use of these public lands: Title 36 of the Federal Code of Regulations 327.2(c) specifically prohibited the use of off-road vehicles on public lands.

In addition to the federal codes and laws being violated at Grey Cliffs, the use of the lands also transgressed some key philosophies and ethical principles supported by the Corps. Edwards stated,

> Well, as a resource manager, my primary mission is to ensure that the public lands, that the public has entrusted with us to [protect], are preserved, maintained for future generations. That's basically our mission. And the secondary component of that is for the public to be safe, when they are using our facilities.

Added to this Corps mission, environmental operating principles also emphasized and encouraged a "low density recreation mindset"; Edwards stressed that this mindset is "not a direction; it's what our principles are; it's how we, you know, trust and take care of our public lands . . . so the mission has objectives, and those objectives are to ensure the public lands are protected for future generations." Clearly, though, Grey Cliffs had strayed far from these principles, mission, and even the law. The published notoriety, through dispatch reports and social media posts, continued to threaten the Corps with what could also be considered documented irresponsibility with possible legal repercussions. Many of these activities had fallen out of alignment with compliance, and Edwards served as a person who could regulate behavior, remedy this problem, and improve the area's reputation. These attempts to regulate behavior stemmed from a moral responsibility, in Edwards's mind:

> So it's kind of a moral responsibility and also a, you know, an ethical one, as well, from an environmental and, you know, recreation perspective. And, and I have a firm belief that no matter what you do in life, as long as you're doing the right thing morally, ethically, legally, um, that it's going to have a good result. And may not be easy for all to understand and see, but, uh, there's a, a, as long as you do it on those principles, then I don't feel like that you can, you know, fail. That's my personal experience.

Edwards's motivations for enforcing compliance, therefore, stemmed from his legal and moral responsibilities, as well as a loyalty to the Corps' missions and goals. Ultimately, these goals were behind Edwards's efforts to persuade the community to protect this area, in whatever form that might take. As Hartelius and Browning (2008) write of organizational leaders such as Edwards, "The manager-as-rhetor notion is a way of understanding how persuasion is part of an organizational leader's role. Any leader—political, spiritual, entrepreneurial—uses language to communicate ideas and direct followers" (p. 28). Edwards, who would not consider himself a formal rhetor, still exhibited an awareness that he needed to convey Corps' values to the community; since the community was not familiar with the Corps except in relation to negative past experiences with land takeovers, it truly did not understand these Corps missions and goals that Edwards needed to communicate, as he attempted to co-construct an ethos and some type of relationship with the community.

In this chapter, I begin by analyzing Edwards's attempt to co-construct his ethos with the community as a regulator. In this role, Edwards attempted to persuade this community to act, based on his role as an authoritative figure within the Corps. I analyze texts that Edwards initially used in this regulatory process, as well as a discussion of the documented resistance of the community, to highlight this failed attempt to reach community members. This analysis reveals Edwards's values as Corps resource manager, many of which contrasted with the community values discussed in the preceding chapter. This lack of aligned values contributed to Edwards's constructing an initial ethos that the community had difficulty accepting.

EDWARDS'S ATTEMPT TO CO-CONSTRUCT THE ETHOS OF REGULATOR: A RHETORICAL ANALYSIS

I used a rhetorical framework for analyzing Edwards's reflective self-narrative (see Table 4.1), including appeals to credibility, such as authority and expertise or experience. Because Edwards was the original organizational orator who began the conversation with the community, and because he was applying the Corps training he had received in the public town hall meetings, I analyzed his ethos appeals specifically to reveal his efforts to address this unique rhetorical situation. The oral data I used to analyze the impact of Edwards's regulatory rhetoric appeared in two main ways: Edwards discussed the impact of his recommendation for closure and its impact on the community during his interview, and I observed community members' reactions to Edwards's presentation during the first town hall meeting. Community members also indicated resistance to Edwards's initial efforts during their interviews. Later, I discuss the ways various texts complemented this regulatory ethos construction. This analysis focuses specifically on the credibility category of Edwards's ethos, specifically his appeals to authority and experience, which Edwards focused on during the first part of the first meeting.

Appeal to Authority

David Edwards's position title indicates authority: Corps resource manager of the X District for the U.S. Army Corps of Engineers. He mentioned this title in the interview that I transcribed, and he began the

Table 4.1. Rhetorical Framework for Analyzing Edwards's Narrative—Credibility

Ethos appeal	Explanation	Example quotations
Credibility		
Authority	Statements of position, such as position title, also statements of job roles and responsibilities.	"So I came to [District X] in 2011, and one of the first things I did as the resource manager, my job is to ensure that the public lands that we've been entrusted with protecting are managed in a way that's conducive to the type of activity and specifically safe for public use." "Well, as a resource manager, my primary mission is to ensure that the public lands, that the public has entrusted with us to protecting, are preserved, maintained for future generations."
Experience or expertise	References to information someone in this position might have access to that others would not, also eyewitness testimony or accounts related to the issue under discussion.	"I have a documented County dispatch report; it's from 2016–2018, I requested that they provide me with, any call that [the] County received in regards to [Grey Cliffs]. Ninety pages of report were given to me." "Syringes and needles that were left on the ground that I found, you know, that was an example, I've seen that more than one time." "[Grey Cliffs] was a dump. I mean, I cannot tell you the amount of trash I picked up there." "And there's a lot of examples out there, too, that I've fortunately been a part of in my experience and career to kind of see, use, and learn from, and, you know, what didn't go good, what went well, what didn't go so well."

© 2021 by the Association for Business Communication. Reprinted by Permission of SAGE Publications

first town hall meeting with the community by stating this title. Edwards also indicated what one of his authoritative roles was within the Corps, which was to enforce Title 36 of the Federal Code of Regulations, which prohibits the kind of environmental and safety problems that had damaged Grey Cliffs so severely. To emphasize this, Edwards states:

> So I came to [District X] in 2011, and one of the first things I did as the resource manager, my job is to ensure that the public lands that we've been entrusted with protecting are managed in a way that's conducive to the type of activity and specifically safe for public use.

In other words, part of Edwards's authority as Corps resource manager for this area was enforcing these regulations. In addition, he also possessed the authority to close the area, and the first meeting began with that option. It was just an option, and Edwards emphasized that he never wanted to close the area, but, because of the problems occurring there, he certainly had the authority to do so, although during the interview he indicated that before making a final decision he was required to involve the public. Working to support the Corps mission in his official role, Edwards strived to protect Grey Cliffs and public safety. Edwards indicated that he was responsible for all 41 access points at this lake, and Grey Cliffs was just one of those points. However, it was the site of the most "nefarious activity." In fact, the Corps had developed a public service announcement and the slogan "Keep your wheels on the street; use your feet" to promote sustainable behavior at all of these points, but, according to Edwards, the Corps took a "[Grey Cliffs] approach" when it created this public service announcement, due to the rapid decline in this area. Using this statement in the narrative also boosted Edwards's authority because it was such a public way of encouraging responsible use of the area. He managed the area that this announcement targeted. All of these statements in the narrative reiterated what Edwards had said at the first meeting, and they were meant to emphasize his authority as Corps resource manager.

Appeal to Experience

Another appeal to credibility Edwards drew upon as he discussed options for solving Grey Cliffs' problems was his appeal to experience. Not only did Edwards appeal to his authority, but he had also read documentation about these problems at Grey Cliffs and witnessed them firsthand. One form of documentation that Edwards cited to bolster his argument that problems had gotten out of hand was the county sheriff reports:

> We need to do something, we . . . have a lot of ATVs, illegal drug use, we have a lot of . . . these aren't instances of just assumption, these are documented cases that the County Sheriff's Office prosecuted, and arrested people based on, what they found through their investigations. And I have a documented County dispatch report; it's from 2016–2018, I requested that they provide me with, any call that [the] County received in regards to [Grey Cliffs]. Ninety pages of report were given to me. And a lot of the reports were domestic related, rape related, drug use related, assault; there was a kidnapping with the individual threatening to murder their son, there was a murder investigation, I mean, there were very significant crimes that were of concern to me and to a lot of the Corps of Engineers leadership because we're, again, public safety's paramount, and if we have an area that has that type of activity, we have to do something to stop it, to deter it, to, you know, change that dynamic.

In addition to these reports, Edwards had also visited Grey Cliffs and had observed some of the problems documented in the reports firsthand. Below are some of the problems he witnessed, which he also emphasized at the meeting to clarify to the community exactly what was taking place:

> Syringes and needles that were left on the ground that I found, you know, that was an example, I've seen that more than one time.

> [Grey Cliffs] was a dump. I mean, I cannot tell you the amount of trash I picked up there.
>
> The trash that we don't see that's under the leaf litter, that, that you step on when you're walking, and just off, you know, the road, it is significant. I mean, the glass bottles, it's all through the bank, it's the worst I've ever seen.

Based on this knowledge, Edwards thought that he could use his experience to improve the area. He had knowledge gained from his position and training as Corps resource manager, but he also had witnessed the problems that had occurred to lesser degrees at other lake-access points in the region. This type of experience would benefit this situation in particular, according to Edwards:

> And then this is where I think my experience, my expertise comes in is that we can lay out a facility that will be sustainable for the environment, and you know, reduce, less runoff and less erosion issues which affects water quality there. And then obviously also increases the type of use and the behavior of people.

Clearly, this appeal to experience and expertise as part of credibility was a recursive process that developed over Edwards's experience in this role that spanned close to a decade. Specifically, after-action reviews, required by the Corps after occurrences such as those at Grey Cliffs, contributed to this recursive development of ethos that would improve future Corps practices by resource managers like Edwards. The whole purpose of these reviews was to allow Corps personnel to reflect in writing upon these types of incidents and determine what went well and what could be improved for future interactions.

Edwards's process of appealing to credibility seems fairly straightforward based on his appeals to authority and experience. After all, his authority could not be disputed as Corps resource manager, and the sheriff's reports and Edwards's personal observations could not be challenged because they had been sufficiently documented. However, the community *did* challenge Edwards's credibility. Instead of accepting these appeals based on experience, they rejected Edwards's testimony about the points he mentioned. While the community had no choice but to accept evidence of the environmental damage that the off-roading had caused because it was in plain view, in interviews community members characterized Edwards's testimony of other problems at Grey Cliffs as "lies." In other words, Edwards's appeals to credibility failed; the community was not receptive to them overall, based on his ethos of credibility alone. This difficulty also manifested itself in the types of regulatory texts Edwards used, which complemented his use of regulatory rhetoric during the town hall meetings.

Regulatory Texts

Edwards employed various forms of texts to persuade the community to change, including maps, warning signs, the sheriff's reports, and stories relating to Edwards's Grey Cliffs experiences. Written texts were also supplemented by Edwards's oral discussion of them during meetings. During the town hall meetings, Edwards was formally given the floor to present some of these texts on stage, and then the community was given an opportunity afterward to ask questions and offer comments. Ultimately, these texts served to regulate the future actions of the community; at least, that was Edwards's purpose in incorporating them. In part, he justified his proposed actions based on these texts. He also used these texts to motivate the community to follow the Corps regulations and preserve the Grey Cliffs area.

Map of the Lake Area

At the first town hall meeting, Edwards was prepared for a somewhat large audience with a projection system, screen, speakers, and microphone. As part of his presentation, he displayed a Corps of Engineers map that highlighted the Grey Cliffs area as a way to provide context for this access point's location when compared to the lake as a whole. From Edwards's perspective, he was informing the audience about the geographic relationship of Grey Cliffs to the other lake-access points he managed. While this map was informative, one thing that seemed to anger the community significantly was the emphasized distance between Grey Cliffs and other lake-access points, which were displayed prominently on the projection screen. While this map was informative from Edwards's perspective, the community viewed it as representative of sanctioned Corps authority that they had failed to negotiate with in the past. This map presented Corps values in ways that seemed to exclude community values and needs; it was a one-sided representation. As Durá (2018) discusses, maps are socially constructed and represent arguments, even though they are based on "factual data"; they "identify how rhetorical choices such as inclusion and exclusion [shape] possibilities of civic, national, historical, and spatial identities" (Eichberger, 2019, p. 14). Maps are story-telling devices, and when conveyed from a monologic perspective, they can contribute to alienating community members who live in the represented areas, since their voices and perspectives may not be reflected in the way the map was originally constructed. In addition, Carlson and Caretta (2021) point out that "maps are fundamentally incomplete, requiring us to consider what elements are left out in their creation" (p. 46), specifically elements that could silence certain voices while privileging authoritative ones (Carlson & Caretta, 2021, p. 46; Eichberger, 2019, p. 18). Many maps, particularly digital ones such as the one Edwards showed on the projection screen, also "limit how communicators can convey a location's topography, history, and local action" (Butts & Jones, 2021, p. 5). Community members pointed out that, according to this map, if Grey Cliffs were closed, they would have to drive 30 minutes or more to reach another access point if they wanted to fish or swim, when some of them lived less than a mile from the convenient Grey Cliffs location. The community valued the proximity and convenience of the area. This map also reflected the ultimate consequences of the community's failure to comply with the Corps rules and regulations: if the area were closed, they would have to travel long distances to continue accessing the lake. This proposed inconvenience was hard for the community to accept, since they had always had local access to these activities; the community strongly voiced their resistance to closure based on this point alone.

Warning Signs, Sheriff's Reports

During the first meeting that provoked the community's hostility, Edwards referred to the Corps' rules that had been broken at Grey Cliffs, citing the 90-page dispatch report from the local sheriff's office. Some of these rules had been displayed on signage physically placed at Grey Cliffs, but Edwards stated that those signs had been vandalized or removed. Newly placed signs had received the same treatment. Edwards narrated:

> Well, we didn't have the support of the community because anything that I'd done previously—I'd put a port-o-jon there; it was burned to the ground. I put signs up; they were shot. Um, you know, we, we put signs up, no ATVs; they were removed; they were shot.

Understandably, Edwards interpreted these actions as lack of community support, and he very well could have extended that lack of support to mean that the community did not care about the area. In Edwards's mind, therefore, the community might not even mind if the area were closed, since they apparently did not care about the area and demonstrated open hostility to the Corps as an organization through Edwards as an individual. His actions originally were intended to show interest in the area and community needs (through providing the port-o-jon, for example), but these efforts were violently destroyed through vandalism. The sheriff's reports that Edwards cited were indisputable evidence of these facts, again supporting appeals to authority and expertise from law enforcement and Code of Federal Regulations perspectives. These reports amounted to "scientific evidence" that Edwards used to bolster his argument, which, while compelling, also hinted that some important information might be missing, such as whether the people committing the crimes were from the local community and present in the audience. Intentional or not, this type of evidence that governmental communicators such as Edwards possess can serve to bolster the arguments of those with access to the information while contributing to the marginalization of those without it (Graham & Lindeman, 2005, p. 443), since, without access to the same evidence, there doesn't seem to be any way to argue against it.

Stories
In the beginning, especially at the first meeting, Edwards was very focused on his own organizational identity story as it related to his own ethos; after all, he held a very high position of responsibility within the Corps. These types of identity stories lend a unique ethos to the storyteller because the narrator can often draw on personal experience in support of the story, and narrating experience lends credibility to the storyteller through details of those experiences; in essence, no one can argue with the details of another's experience as expressed by the one who experienced them. These experiential details become facts that help support arguments the narrator promotes along a diverse gamut of formal and informal storytelling. Specifically, Edwards had told of times he had been to Grey Cliffs and found needles there, indicating drug use. He described vandalism of the area, as well. A large portion of the stories, supported by the sheriff's reports, revolved around crimes committed in the area. Complicating these stories were ones Edwards told about unauthorized off-roading in the area. Edwards could observe the environmental effects of this off-roading, such as erosion, but he also heard and told of stories from community members such as those who had experienced trespassing, as off-roaders accessed people's land that surrounded the Corps' property. These stories were essentially oral texts presented at the meetings and during my interview with Edwards, heightening the severity of the problem and validating the organizational Corps identity as a regulating authority, with Edwards as its sanctioned representative and conveyor of a master narrative.

Edwards's appeals to credibility, specifically to authority and expertise, allowed Edwards to construct a regulatory ethos that these various texts supported and documented. Edwards's rhetorical appeals and texts also supported a dominant Corps narrative, one that did not include the community. Rather than persuade the community, Edwards found himself in strong opposition to his community audience, an audience who was unwilling to acknowledge the authoritative ethos he created.

THE COMMUNITY'S INITIAL RESPONSE TO EDWARDS'S REGULATORY RHETORIC

Before the first town hall meeting, rumors began circulating in this small community that the area would indeed be closed, and this possibility generated a lot of publicity ahead of the meeting. About 200 community members, local government leaders, the media, and a congresswoman's personal representative were in attendance. The atmosphere was lively and tense as Norma began the meeting, then introduced Edwards. Edwards began by emphasizing the problems now endemic to the area and stressing that now was the time to do something to preserve this area before the crime and environmental damage got even further out of hand.

During this presentation, Edwards was clearly the authority on matters relating to Grey Cliffs. He narrated the crime statistics and costs the area had generated over the past couple of years, and he displayed the map. After presenting his evidence, he then stated that closure was one option to allow the area to repair and become off limits to all, including criminals and would-be criminals. As he spoke, Edwards commanded the group well and sounded knowledgeable about the area's activities and problems. However, even though Edwards clearly seemed to operate from a position of strong agency within this context, the community instantly reacted with hostility to this message of possible solutions, with closure being at the top of the list. The vocal crowd responded with a *polyphony* of voices (Bakhtin, 1984; Bondi & Yu, 2019; Castelló et al., 2013), each demonstrating various reasons why the area should not be closed, although a few expressed agreement with the idea of closure. In this sense, the community's voice was also functioning in *symphony* (Bondi & Yu, 2019), since most were in opposition to closure—although for different reasons, one of which could have been that no one understood what benefit the community would receive from the closure. In such cases where the connection between risk and benefit is not clear (Simmons, 2007), "the public response is predictably negative, and knowledge-gathering activities will tend to focus on amassing evidence to support a foregone conclusion: that the risk is unacceptable" (Tillery, 2019, p. 11). In addition, Edwards was up against poor public perception that can sometimes accompany the role of government organizations in public life, inhibiting the construction of an ethos of goodwill and wisdom (Tillery, 2006, p. 326). This type of reaction can reflect skepticism that community input will truly have an impact on final resolutions to environmental issues. In the Grey Cliffs case, the risk of closing the area for rejuvenation was not worth future benefits the community might receive; the community feared the area would not reopen.

Community members were not persuaded by the Corps stories that Edwards narrated, which portrayed their use of the area in a negative light; rather than being motivated by the stories to comply with the regulatory action Edwards proposed, they said the stories were lies and offered their own contrasting, positive stories as a rejection of Edwards and the changes he proposed. Their stories of the area were different and supported their cultural views (focusing on Grey Cliffs' virtues); their personal experience narratives were in complete opposition to the Corps' stories, overarching narratives, and its values that Edwards's stories reinforced. This particular community valued its own stories (see Chapter 3), not only about their personal family experiences at Grey Cliffs but also about the established negative narratives of Corps involvement over the years.

During the first town hall meeting and afterward, many community members began questioning Edwards further about the drug use and crime statistics, stating that they had never seen Grey Cliffs used for such nefarious activity. They protested that the community had not been asked how it could help and that they had plenty of resources to maintain the area; after all, one community member's father had actually paved the road that led down to the lake-access point many years ago. The community argued that since they were all living near the area, they could informally police, help maintain, and monitor it. Their passion and instant rejection of Edwards's ideas were a shock, especially to Edwards, who termed this meeting, in part, a "failure": "I thought I failed . . . like the first meeting, I thought I was pretty open that I don't want to close the area; I don't want to close the area, I said that five or six times." Mackiewicz (2010) discusses how sometimes this type of failure can occur because the audience believes the communicator may have something personal to gain from the changes proposed (pp. 407–408). The knowledge that the speaker has may be eclipsed by an assumption of personal gain, and the audience therefore does not trust the speaker. Given the tense past relationship between the Corps and the community due to the land buyouts, the audience very well could have been suspicious of Edwards as they rejected the "evidence" of his credibility presented to the community. Mackiewicz (2010) emphasizes the dialogic nature of the development of credibility and that sometimes this credibility can actually be deconstructed between communicators and audiences (p. 408). In situations like this, even though a rhetor may be attempting to construct a credible ethos based on experience and authority, an audience can reject that attempted construction. As Eubanks (2015) clarifies, "Enculturation into a group is what allows us to construe what is credible and what is worthy of attention" (p. 121). Because Edwards did not demonstrate "share[ed] values and practices with others in [his] . . . social [group]" (Eubanks, 2015, p. 121), his first efforts to communicate with the community were not successful. He did not indicate at all that he could even be part of the community's "social group" as he focused on credibility while communicating with the community about Grey Cliffs' issues.

As part of this authoritative role, using the "language of technical functionalism" (Patterson & Lee, 1997, p. 29), Edwards had narrated the facts of environmental degradation and crime statistics to the community audience, not really realizing the impact this narrative was having on those in attendance. Not prepared with alternate statistics or facts other than their own antenarratives and counterstories focusing on Grey Cliffs' virtues, community members struggled against Edwards's negative portrayal of them and their behavior toward Grey Cliffs. Community members, drawing upon their own values that differed markedly from Edwards's, contributed to a growing "confirmation bias" (Eubanks, 2015, p. 54) among themselves that Edwards simply could not be right about the statistical information or experiences he presented; Edwards certainly could not be trusted. Not willing, initially, to accept this negative narrative to any degree, community members totally rejected Edwards and his message.

Another reason the community may have initially rejected Edwards is the growing level of citizen involvement in scientific and environmental issues impacting the general public (Tillery, 2019, p. 1), in which scientific expertise is often challenged. Lindeman (2013) discusses such a trend, especially within discussions of environmental conflicts. Part of the reason for such challenges could originate from the fact that "environmental problems, due to their complexity, often elude precise or conclusive solutions" (Lindeman, 2013, p. 64), therefore opening the door to more challenges through "democratized expertise" (Lindeman, 2013,

p. 65; Nowotny, 2003). However, as Tillery (2019) points out, "scientific discourses and environmental discourses have been inextricable" (p. 6) throughout the history of American environmentalism, and this connection has created some tension for those not trained in the sciences, who nevertheless would like to participate in environmental discussions and policy decisions (Simmons, 2007). Citizens now expect to be involved in decisions that impact them and "expect experts to be responsive to their concerns" (Lindeman, 2013, p. 65), even to the point of community members' viewing these experts as less credible because of their "institutional affiliation" (p. 81). The Grey Cliffs community very well could have been responding with these same types of expectations in mind.

The Grey Cliffs conflict exemplifies how two parties, the Corps and community, both had good intentions: they both wanted to be involved in keeping Grey Cliffs open. However, Edwards had to struggle against the negative narrative the community had preconstructed about him as a member of the Corps, and the community fought to resist the negative narrative that Edwards told of it, including the crime statistics assigned to it, which would be hard to dispute. While everyone also agreed that these two main parties conflicted (the discussion was so hostile that no one could disagree with this point), Edwards realized that he could not encourage positive change if the community was not on board with him. Something would need to change so that the community would want to begin collaborating with him. Edwards needed to negotiate agency, ethos, and the coinciding social action with the community, effectively co-constructing an ethos and relationship with it. Although Edwards portrayed governmental authority, and his apparent "agentive space" (Herndl & Licona, 2007, p. 143) was also implied, based on his position, he realized successful social negotiation with the community was an essential prerequisite for this process to even begin.

While Edwards had the freedom and authority to present an identity he thought was appropriate, the community did not reflect that identity back to him (Hyland, 2012); the community did not accept the discourse Edwards promoted. Instead, it reacted with strong disapproval of Edwards's ideas. The community rejected the identity Edwards had assigned to them by insisting that those events hadn't happened and that they had never observed such activity in the area. While the community was eventually willing to accept the transgression of off-road vehicle use as part of their identity, perhaps in part because its effects were so obviously detrimental, they were unwilling to accept the other characterizations. The more the community was presented with Edwards's characterization of them, the group ethos Edwards had assigned to them, the more blameworthy the community seemed, and the further all parties appeared to be from the possibility of negotiating a resolution. Denial seemed to be an easier choice for the community.

At this point, Edwards and the community members did not share enough common values to trust each other; these differing values contributed to different identities that Edwards and the community were portraying, and these identities were not aligning. Edwards recognized this lack of trust himself:

> And so, I knew going in [to the town hall meetings] that there was some anti-government [sentiment] and wanting to take the area over individually, and there was a lot of different pushes to, uh, self-govern. I heard that a lot and still hear it.

This distrust of government authority also contributed to the community's lack of acceptance that Grey Cliffs truly did have some environmental problems and evidence of criminal activity. While the community could view the mudslides, trash, evidence of hunting/shooting damage, and runoff into the lake, they never really did accept the crime statistics that Edwards cited as a reason for closing the area. In his interview, Tom vehemently stated a few different times his belief that this information about criminal activity was "bare faced lies": "I called him [Edwards] out, you may know, you may have remembered, you know; I called him out on a few things; and I told him I just thought it was bare faced lies." Regarding the needles themselves, which, to Edwards, were evidence of drug use, Tom narrated:

> We may have picked up a few things, but nothing like. And all this about needles and stuff—I've never seen or never picked no needles up.

> Hey, you know, I don't believe none of that, 'cause. Yeah, 'cause I was the one told him, I said, I's overseas, so I don't know how they's seeing needles and everything down there; I'm not, you know.

Tom believed that his own military experience, in part, would help him identify evidence of this activity if it were truly going on. Hartelius and Browning (2008) state, "When the audience and the rhetor share a cultural identity, they can communicate without making every assumption explicit" (p. 24). This shared identity was missing, along with common values. As a result, negotiated, social action was also missing at this point.

In this current stalemate position, Edwards and the community had to renegotiate to develop identities and a collective ethos that seemed suitable for a relationship between these two parties. A new or revised narrative also needed to be written, one that made provision for a negotiation with the area's cultural history and context (Faber, 2002; Scott et al., 2006; Zachry & Thralls, 2007), an acknowledgment of how important this area was to the sustainability of this community.

Another active community participant and organizer, Norma, did not achieve acceptance from the community either, even though at the outset, it seemed she would, based on apparent shared values and ethos. While she desired to negotiate ethos with this community (in many ways, *her* community), many obstacles stood in the way of the community's accepting her as a leader and organizer who could help them. Chapter 5 analyzes ways Norma's attempted construction of ethos, values, and agency were impacted by community rumors and distrust, despite the community view that Norma could very well be the needed instigator for change in this crisis. The community's interactions with Norma illustrate the need for trust as a jointly shared, negotiated emotion when beginning to resolve conflict.

KEY RECOMMENDATIONS FOR TECHNICAL, PROFESSIONAL, AND ORGANIZATIONAL COMMUNICATION AUDIENCES

- Realize that some audiences may not respond to rhetorical appeals that other audiences might respond well to.
- Because appeals to credibility and authority alone may not be effective, especially for community audiences who might be considered marginalized or disadvantaged, identify rhetorical appeals that resonate with these unique audiences.
- Be aware that texts not incorporating community values and that reflect the dominant, opposing narrative may be rejected by the community.
- When designing texts impacting local communities, ensure community members can contribute to, understand, and identify with these documents, providing them with meaningful participation and control.
- Share ownership of document creation with local community members to demonstrate collaborative ownership and investment.
- Incorporate language and values that local communities understand. Even genres normally accepted by "traditional" audiences, such as the warning signs with regulations, may not be accepted if they originate from language and values foreign to local communities.

KEY RECOMMENDATIONS FOR ENVIRONMENTAL SCIENCE AND PUBLIC POLICY COMMUNICATION AUDIENCES

- Identify and address potential political and cultural implications when applying scientific or other evidence to creating public policy.

- Minimize community resistance by incorporating the interests and values of impacted communities when creating, discussing, and evaluating policies conveying scientific and environmental regulations.
- Listen to seemingly negative stories that rebel against dominant narratives to identify ways community members may be experiencing oppression given the community's cultural history.
- Look for clues to community members' values, thoughts, and feelings behind potentially problematic actions, revealed in what appear to be negative community stories and hostile responses.
- Include community members in communicating about and developing documents and policies related to their local environments.
- Acknowledge that local communities may have difficulty understanding connections between policies and regulations and their own behavior when policies and regulations conflict with "standard" acceptable behavior among community members. As a result, communities may be inclined to "just do as we've always done."

5

ATTEMPTING TO PERSUADE AS A COMMUNITY ORGANIZER
Norma's Narrative of Logic Without Emotion

> *"Some situations do not have 'happy endings,' nor would a happy ending be the realistic outcome. Some situations need to have matter-of-fact communication to function. Period. I am a scientist. I do not think with my limbic system."*
>
> —Norma, community organizer for meetings related to the future of Grey Cliffs

Norma organized the first few town hall meetings, during which the Grey Cliffs closure was proposed and discussed. During the first part of this conflict when rumors began circulating about possible closure, Norma ran into David Edwards, the U.S. Army Corps of Engineers resource manager, at the Grey Cliffs lake-access area and began talking with him about the crime that had been occurring in the area. One personal concern Norma voiced during her interview was that "closing Grey Cliffs would cause the criminal element to seek other out-of-the-way places to do their 'business'; the neighborhood already had crime problems, and closing Grey Cliffs would push crime into my, and other neighbors', properties." She also mentioned that she swam for exercise, and if the area were inaccessible to the public, she could not drive long distances to continue doing that. According to Norma, she heard about the possible closure "among neighbors, sheriff's deputies, [and] county road crews. Issues were not rumored, only that *closure would* happen." As Norma and the neighbors continued discussing the issues, they mentioned to Edwards the idea of holding a community meeting to inform others of what was going on and help determine how to resolve the growing problems of crime, trash dumping, and other environmental damage in the area. Inevitably, the option of closure would also be a topic discussed at the town hall meeting.

https://doi.org/10.7330/9781646425761.c005

Based on her involvement with nonprofit organizations and grant work in the past, as well as her vested interest in this issue, Norma took the initiative to organize the first community meeting to discuss the Grey Cliffs lake-access closure. Her personal interest as a member of this community, as well as her dedication to help neighbors with similar concerns, combined to generate the motivation Norma needed to pursue this work. She went door to door and talked with community members about the need for a meeting; she created fliers advertising the meeting date, time, and place to tape outside neighbors' mailboxes and post at the meeting location; and she initiated phone calls to recruit government officials and other representatives, such as the local media, to attend. She also prepared for the first meeting by creating an agenda and planning how to begin the meeting, including the presentation Edwards made. As a community leader, Norma spent a large amount of time and effort to make the first town hall meetings a reality.

During this first meeting in particular, Norma stood out as a powerful force that would organize this community for action. Norma's goal during the Grey Cliffs conflict was to begin an organization, a nonprofit organization that could legally and officially partner with the Corps to solve the issues of crime and environmental abuse. This option seemed like a logical opportunity so that the community could interact with the Corps in an official way that would be legally binding. In relation to Norma, community members were working together with a common goal of keeping the area open; they saw working within the organizational framework that Norma started as a way to begin collaborating toward solving the problems facing Grey Cliffs that prompted possible closure.

This chapter analyzes ways Norma's attempts to negotiate an ethos with this community were impacted by her lack of character appeals, such as those based on affinity and sincerity, as she presented her own organizational narrative, which was grounded in grassroots efforts and knowledge. Rather than attempting to make connections with this community as a whole, Norma relied only on her ethos based on past work experience and credibility she had gained from that work experience, similar to what Edwards had done. This rhetorical analysis reveals Norma's efforts to construct a credible ethos, based on experience and expertise as well as several of Norma's own apparent values, evidenced through this ethos construction. The grounded theory analysis also identifies community responses to Norma, resulting from her reliance on credibility alone: resistance based on her being overly controlling of the information, lack of personal connection with the community, and untrustworthiness. This discussion also includes texts that Norma incorporated into her communication efforts that revealed values she hoped would motivate the community to act. Ultimately, this chapter serves as a case study of a communicator who achieved success through some of her persuasive efforts but did not attend to some crucial character appeals that ultimately resulted in the community's rejection of her and her message. This case study reveals the necessity of incorporating and negotiating these missing ethos appeals as well as trust building, especially when attempting to address and work with a hostile audience. These ethos appeals and trust building relate to negotiating with community members using an aligned values perspective as well.

NORMA'S ETHOS: CREDIBILITY, EXPERTISE, AND EXPERIENCE

Norma presented a strong persona at the first community meeting; her individual ethos seemed strong (Aristotle, ca. 367–347, 335–323 B.C.E./2019). Not only did she seem credible based on her past work experience but also was extroverted, took control, and stood at the front of the large group of about 200 people with a flip chart so that she could take notes based on ideas presented. She facilitated the meeting, took questions, and assigned questions to Edwards or other officials attending to answer. She moderated comments when they became too hostile. She demonstrated experience keeping the meeting focused and managing a difficult crowd, and she portrayed an ethos of confidence and credibility. The community needed this type of leader to help them during this time of emotional crisis.

When I interviewed Norma about her role as a community organizer, she described herself as having various skills, background knowledge, and experience that could assist the community during this conflict. These self-characterizations amounted to the individual ethos Norma attempted to present to the community: her qualifications based on credibility, expertise, and experience, as shown in Table 5.1. Quotations below are Norma's own words gathered during my interview with her; due to health-related reasons, the interview was conducted via email, and, therefore, the text is reproduced as she composed it, including all abbreviations and punctuation.

From all outward appearances, Norma was the perfect candidate to fully function as an effective community organizer based on her self-described qualifications. Perhaps to enhance her qualifications even more and build on antigovernment sentiment already prevalent in the

Table 5.1. Rhetorical Framework for Analyzing Norma's Narrative—Credibility

Ethos appeal	Explanation	Example quotations
Credibility		
Authority	Statements of position, such as role as organizer and accomplishments in that role, also statements of previous job roles and responsibilities.	"I am a scientist." As a community organizer, Norma "envisioned forming a lasting organization, . . . that it should be of a 'legal' structure in preparation for incorporating into a legal entity, at some point." Norma "made inquiries with accountants, attorneys, business assistants, etc., about forming a 501(c)(3)" and created "guidelines published on Facebook to be used at the first committee volunteers meeting." She possessed prior knowledge of "the importance of . . . organizational structure being formed [and] the written notes required." Norma knew that "only [a nonprofit organization] can enter into a legal agreement with the Corps of Engineers." Her goal was to organize the community "to become a functioning 'organization,' to set goals, and execute plans."
Experience or expertise	References to information someone in this position might have access to, based on personal lived experience, that others would not, also eyewitness testimony or accounts related to the issue under discussion.	"When I happened to meet [David Edwards] at Grey Cliffs, [I] was given facts about Corps of Engineers' intentions during an hour ± long chat." She "lived in and used the area for 30+ years."

community, she also distributed Convention of States literature at the meeting. Knowing that the area was generally conservative politically, Norma began building ties with this community by letting them know right away that she was in the same corner politically, through her association with the Convention of States, a group that tries to minimize federal government control and would provide for the passage of constitutional amendments, if two-thirds of state legislatures propose them through a Convention rather than having Congress and Senate amendment proposals and votes alone. This attempted connection with the audience, which could be interpreted as an appeal to affinity, appeared to be very subtle, and Norma did not mention the literature or her connection with the Convention of States at all during the meetings. She relied only on her prior distribution of this literature to make this point.

NORMA'S CONTINUED ETHOS DEVELOPMENT: COMMUNICATING VALUES TO MOTIVATE ACTION

In contrast to David Edwards, who represented Corps values that appeared on the Corps' website and its guiding organizing principles, Norma did not have such websites or organization-sanctioned documents to use as sources for her values and organizational texts other than the Convention of States literature that she distributed at the first meeting. Nevertheless, based on her interview with me, fieldwork observations, and interviews with community members, I found that Norma did communicate specific values and reinforced those values through organizing texts. These values and texts facilitated the action of the community to some degree, especially at first.

Because language is sermonic (Weaver, 1970) and communicates values, language use can motivate dialogue, persuade listeners to accept values, and encourage audience members to mobilize and act in ways that align with those values. In her role as community organizer, Norma communicated several values that were also carried through in the texts she used. These values were the importance of face-to-face communication, her belief in focusing just on facts, her rejection of emotion when discussing Grey Cliffs, consistency in carrying out original plans through defined organizational structure, and her belief in grassroots efforts. Norma's success in her role as community organizer and motivator ultimately would be determined by how well she could persuade the community to accept these same values. I present these values, along with their corresponding evidence, in Table 5.2.

Importance of Face-to-Face Communication

Norma believed in face-to-face communication. She had first heard about the closure possibility when speaking in person with David Edwards at Grey Cliffs. At that time, she suggested to Edwards that a town hall meeting with the community would be appropriate so that everyone could discuss the situation and their concerns. Following Edwards's acceptance of this idea, Norma organized the first town hall meeting. Then, she went door-to-door, visiting with neighbors, letting them know of Grey Cliffs' status, and motivating them to attend the first town hall meeting. Norma's recruitment efforts were a huge success given the large group from this small community who attended the first meeting. Norma verified that success by acknowledging "such a large turn-out for public forums" during her email interview. At the first town hall meeting, Norma was clearly the leader, since she began the meeting,

Table 5.2. Coding Scheme for Norma's Values

Code (value)	Explanation	Example quotations / Other evidence gained through fieldwork
Importance of face-to-face communication	Statements or other observed evidence that Norma valued being physically present when addressing this conflict with the community.	Stating "such a large turn-out for public forums" as evidence of success.
		Organizing the first town hall meeting. Visiting neighbors door-to-door and distributing fliers.
Belief in focusing on facts	Statements using the words "logic" or "fact."	Norma "attempt[ed] to learn facts" in response to hearing that Grey Cliffs might be closed. "When I happened to meet [David Edwards] at [Grey Cliffs], [I] was given facts about Corps of Engineers' intentions during an hour ± long chat." "I identified my opinions as my own, backed up their formation by stating facts." "Ideas were logical and fact based."
Rejection of emotion	Statements indicating that emotion was not important, not logical.	"I am a scientist. I do not think with my limbic system." " 'Feelings' are irrelevant." Community members not accepting leadership "either could not critically think, did not think through 'cause and effect,' or thought solely with emotion." "Emotion should not play any role in determining [organizational] communication." "Running any organization on emotion is ruinous to the goals of the business, which must be based on reality instead of 'feelings . . .'" "There is no . . . 'emotion' tied to forming an NPO [nonprofit organization]."

continued on next page

took notes on the flip chart, introduced Edwards when he spoke, and moderated questions afterward. For Norma, these face-to-face communication opportunities were times to influence community members to accomplish her valued goals.

Belief in Focusing on Facts

At the beginning of the Grey Cliffs conflict, Norma believed "that *closure would* happen," despite the Corps resource manager's later insistence that closure was only one of several options to consider. Based on this assumption, Norma "attempt[ed] to learn facts" in response to this knowledge. She also emphasized that "when I happened to meet [David Edwards] at [Grey Cliffs], [I] was given facts about Corps of Engineers' intentions during an hour ± long chat." She also reflects on the role of her personal opinion as she led the meetings: "I identified my opinions

Table 5.2—continued

Code (value)	Explanation	Example quotations / Other evidence gained through fieldwork
Consistency in carrying out original plans through defined organizational structure	References to forming a nonprofit organization that would be a liaison with the Corps, statements about how this organization should be run or how the town hall meetings should be organized.	The goal of solving the conflict was "forming a lasting organization . . . that . . . should be of a 'legal' structure in preparation for incorporating into a legal entity, at some point." "My written suggestions were not followed because the members did not understand, nor seek to understand, the importance of the organizational structure being formed, or the written notes required." "There are rules and formalities established and followed within any organization. You can't just show up and do what you want, when you want to." "People who volunteered for committees did not follow my written guidelines, nor were they familiar enough with 'business formation' to see the importance of the 'process.' Therefore, nothing moved forward." Conducting research online about exactly what was needed to form a nonprofit organization within this state.
Grassroots efforts	Statements indicating that efforts to foster change with governmental organizations should begin with interested and motivated community members.	"People were needed who understand what is required—or are willing to learn—to become a functioning 'organization,' to set goals, and execute plans." Distributing Convention of States literature at the first town hall meeting. Encouraging the community to ultimately take control of resolving this conflict, such as by communicating with the XYZ Company through a developed "elevator pitch."

as my own, backed up their formation by stating facts." The solutions Norma identified, such as establishing a nonprofit organization to enter into an official agreement with the Corps, were also "part of my opinions, which I had identified as such." In addition, according to Norma, her "ideas were logical and fact based." This focus on logic/logos, based on her experience and expertise, formed the basis of her interaction with this community. Her communication efforts, though, did not include any consistent, significant emotional appeals to connect with community values as a whole. In fact, Norma stated that emotion had no place in her communication efforts.

Rejection of Emotion in Addressing Grey Cliffs

The sole focus on logic inevitably resulted in a lack of appeals to emotion as Norma communicated facts to the community. During her interview, Norma stated, "I am a scientist. I do not think with my limbic system." Throughout the interview, Norma made several references to problems with focusing on emotion in communication. This dissociation from emotion, gained in part through her disciplinary training as a scientist, spread to her role as town hall meeting facilitator and community organizer. The tension between Norma's values based on logic and facts and the community's affective values based on strong personal connection to Grey Cliffs and its virtues was striking: in contrast to the stories community members told about their connections to the area, Norma indicated that "'Feelings' are irrelevant." Those community members not accepting her leadership, plans, and recommendations based on Norma's opinions "either could not critically think, did not think through 'cause and effect,' or thought solely with emotion." As Norma defined it, "Emotion should not play any role in determining [organizational] communication."

Norma continued reflecting that incorporating emotional appeals was irrelevant, since the community cannot be considered an organization. Instead, it "[was] more akin to a church congregation" in the diversity of its members, present for different reasons, goals, and purposes, but for the most part desirous of keeping the area open. Recognizing the audience as similar to a church congregation indicates Norma was aware of the need for a different type of persuasion to motivate the audience, one that could include emotion, as many sermons do, yet the focus in the town hall meetings remained on logic and facts and the need to "separate logic from emotion." This separation "would avert problems caused by considering everyone's 'feelings.'" After all, "there is no... 'emotion' tied to forming an NPO [nonprofit organization]," and forming a nonprofit organization was Norma's ultimate goal. The community's focus on Grey Cliffs' virtues, communicated through stories, was, therefore, irrelevant.

Consistency in Carrying out Original Plans Through Defined Organizational Structure

Norma's original goal in solving the Grey Cliffs conflict was "forming a lasting organization... that... should be of a 'legal' structure in preparation for incorporating into a legal entity, at some point." Creating this nonprofit organization was necessary because "ONLY an NPO can enter into a legal agreement with the Corps of Engineers." Norma maintained this goal consistently throughout her interaction with the community. Part of acting out this goal involved conducting research online about exactly what was needed to form a nonprofit organization within this state and then conveying those instructions to the community members. However, Norma found the community's response to these instructions frustrating: "It's become obvious that the neighbors are not interested in pursuing this; that the neighbors are satisfied with as much progress as has been made [regarding] the current outcome." The community had decided that it did not want to pursue creating a nonprofit organization to collaborate with the Corps. This community decision was frustrating to Norma given all the work she had accomplished on behalf of starting this nonprofit organization so far. She stated:

> At the critical development time, when I was absent, my written suggestions were not followed because the members did not understand, nor seek to understand, the importance of the organizational structure being formed, or the written notes required. They took off on a tangent which has stopped the process forward.

When I questioned Norma further about the community's lack of response to her plan of forming the organization, she stated that "people who volunteered for committees did not follow my written guidelines, nor were they familiar enough with 'business formation' to see the importance of the 'process.' Therefore, nothing moved forward. So, back to square one." For Norma, a sharp disconnect existed between the success of her first efforts at motivating this community to discuss these issues at the town hall meetings and the failure to move the community forward in forming this "lasting organization" through creating a jointly constructed ethos with the community. This co-constructed ethos was a prerequisite for collaborative social action as well as relationship. Not willing to see future community compliance with Corps regulations as a possible step forward or perhaps the accomplishment of current sensemaking, she resigned herself to the following:

> Communications restarting with the neighborhood *from the beginning* may put efforts back on track to accomplishing long-term goals. I will have to canvas the neighborhood on foot again, and since people who I had just met last fall will remember me now, I hope to put together a neighborhood directory of "contacts" to make communication less labor-intensive.

Norma did not specify exactly how communications might be restarted in a way that would yield different results based on this community resistance, other than restarting the communication/recruitment process.

Grassroots Efforts

In her interview with me, Norma did not include the word "grassroots." However, she did mention it in a flier describing the type of information needed during one of the town hall meetings, "grassroots input."

That description, coupled with the Convention of States mindset of states' independent authority that Norma indirectly promoted through displaying the Convention of States literature, encourages community presentation of ideas. The *Encyclopedia Britannica* defines grassroots as a

> type of movement or campaign that attempts to mobilize individuals to take some action to influence an outcome, often of a political nature. In practice, grassroots efforts typically come in two types: (1) efforts to mobilize individuals either to turn out to vote or to vote a certain way in an upcoming election and (2) efforts to mobilize individuals to contact a policymaker or other individual with influence to take a particular action (also called "outside lobbying"). The distinguishing features of grassroots movements or campaigns are that (1) they mobilize masses to participate in politics (such as contacting their legislator or turning out to vote) or some other cause and (2) they are conducted through narrow communications such as mail, e-mail, phone calls, or face-to-face visits rather than broadcast media such as television or radio. (Bergan, n.d.)

Some notable similarities between this definition and Norma's role are the "efforts to mobilize individuals to contact a policymaker or other individual with influence to take a particular action" and that these efforts "are conducted through narrow communications such as mail, e-mail, phone calls, or face-to-face visits." Norma enacted these same types of communication with the community and government officials; she also indicated that people were needed "who understand what is required—or are willing to learn—to become a functioning 'organization,' to set goals, and execute plans." Even though Norma did have a clear idea of what her values were and the result she hoped to see through the development of a nonprofit organization that would act on behalf of Grey Cliffs, she still wanted the community to take ownership of these values and goals and take control of the steps needed to ensure long-term goals were met. Fostering this growth through generating grassroots efforts was one way to accomplish this process.

Despite this extensive background, experience, and supporting values, though, Norma had some difficulty connecting with the community and motivating it to follow the rules that she had developed for forming an official organization. Norma attempted to regulate this community's social action and agency for change by presenting a clear, logical plan to follow for moving forward, framed by rules she had gathered through her own experience and personal research. Her focus was clearly on ethos appeals to credibility in her leadership narrative, which resulted in rules, order, and predictable, expected outcomes that she presented to the community.

CHALLENGING NORMA'S ETHOS: COUNTERSTORIES FROM THE COMMUNITY

Reflecting this focus on logic and the need to gather facts during the early part of her involvement, Norma had typed up lengthy notes about the problems at Grey Cliffs, possible solutions, input from the meetings, and dynamics involved in changes to be accomplished and resources required. She planned to use these notes in future nonprofit organization creation. However, despite her experience and efforts to help, rumors began circulating among community members soon after the meetings started and inhibited Norma's developing ethos in its infancy before she could even begin co-constructing it with the community. These rumors and stories served as part of a larger antenarrative similar to what had been constructed about Edwards and centered on several themes of her character: that she

- was overly controlling of the information
- had a lack of personal connection with the community
- was untrustworthy

Some of these themes overlapped to some degree, and community participants offered several examples during their interviews that fit easily within these three categories. Based on the community members' interviews I transcribed, I identified these themes, using grounded theory and interpretive approaches applied in other parts of this case study. By using these approaches, I hoped to study these participants' self-narratives and learn more about these community members' "distinctive versions of their selves" (Boussebaa & Brown, 2017, p. 22) as revealed in their own words. These self-narratives offered unique perspectives on the community's deconstruction of Norma's ethos.

Overly Controlling of Information

From the onset of the conflict, Norma clearly represented the ethos of an informed community member who was familiar with grassroots organization efforts. She distributed Convention of States literature at the first meeting, took charge of the meeting, and, during the second meeting, encouraged meeting participants to organize their efforts further by developing a nonprofit organization that could legally collaborate with the Corps through an official agreement. Despite her authoritative and logical persona, however, rumors and discontent developed surrounding her control of information.

Controlling the Meeting Agenda and Dictating What Officials Said at Meetings

At the first meeting, David Edwards presented the possibility of closing the Grey Cliffs lake-access area as one solution for taming the crime and environmental damage that had been occurring. Because rumors had already been spreading among the community about this possibility before the meeting occurred, Edwards's statements prompted this community to believe that their worst fears were being realized. Even though Norma did not publicly advocate closure herself during the first meeting, rumors circulated that she had coached Edwards and other officials about what to say at the meeting to move toward closure. According to Tom, an active community participant, Norma played a prominent role in regulating the conversation about Grey Cliffs: "But we was told that [Norma] had told him [Edwards] up front everything to say, and she wouldn't let him answer questions out here because she took the microphone, remember?" Denise, Tom's wife, supported Tom's statement: "She told [Edwards] everything to say, and she told the sheriff [Bob Wheeler] everything to say." These suspicions could have been just that, suspicions, but Tom elaborated:

> Do you remember when [Bob Wheeler] said somebody asked him something, and he said, "I can't answer 'cause she told me not to." He throwed her under the bus: "She told me not to answer that. She told me to set here and just be quiet." He throwed her under the bus, and I'm like, why would she do that, you know?

According to Tom, even Edwards himself seemed to acknowledge he had been coached on what to say before the first meeting:

> He [Edwards] said they was a lot of words put in my mouth that, that shouldn't have been spoken. He, all he would say is, "You know who I'm talking about." And I said, "Norma?" And he's like, "You know!" And he'd grin and say, "You know who I'm talking about."

Based on what these community members had heard about the ways Norma was trying to control what these officials said to influence closure of Grey Cliffs, the community also grew to suspect Norma's motivation for organizing the meetings, to some degree; while she did solicit ideas and feedback from the meeting attendees, the focus of the first meeting necessarily remained on the problems occurring at Grey Cliffs. Toward the end of the first meeting, attendees began talking about possible solutions to the problems, but many were in denial that the problems existed, since they had not witnessed some of this criminal activity before and distrusted Edwards's perspective. This lack of belief contributed to the suspicions surrounding Norma and the government officials, all of whom agreed that the criminal and environmental degradation of the area had escalated beyond control. The only apparent solution, at least at first, seemed to be to close the area down.

Attempting to Control the Conversation With a Local Business in Support of Cleanup Efforts

At a later point, Edwards had recruited Tom and another community member, Dan, to visit the XYZ Company to request funds and materials for cleanup efforts at Grey Cliffs. Norma was not part of this visit, perhaps because of a chronic illness she had mentioned during her interview that in part kept her from later meeting involvement.

Before Tom's visit with the company, though, Norma contacted him by phone to attempt to coach him on developing an "elevator pitch" in an apparent attempt to coach Tom and perhaps control the conversation from afar. Tom expressed his frustration with Norma as he narrated:

> Now get this, [Norma], she kept texting me, you know, hey, we need to talk, we need to talk. I told [Denise], I said, I do not want to talk to that woman, you know, and I just got pushing to ignore her. So she finally, she calls me that morning before I'm supposed to leave, uh, [Dan] was supposed to pick me up. Me and him was going over to [XYZ]. She says, "[Tom], what's your game plan?" I said, "[Norma], what do you mean my game plan?" "Well, throw me your elevator pitch."

According to Denise, the purpose of this rehearsal was to allow Norma an opportunity to approve and critique Tom's presentation beforehand: "Yeah, see if I listen, believe in it," she said in reference to Norma's anticipated response. However, Tom did not have any intention of sharing his elevator pitch because he didn't have one. Furthermore, Tom didn't seem to think he needed one:

> And she's like, "Well, just tell me what you're going to tell them." "[Norma]. I don't know what I'm going to tell them because I don't know what we're going over there for. If I knew what we was going for, I would tell you what I'm going to tell them. But we're just going over for a meeting; they evidently want to put in help; they've got some funds, and they want to help clean up."

Clearly, Tom was frustrated with Norma at this point for trying to shape the elevator pitch before the visit to the XYZ Company; Tom also did not seem to be familiar with the elevator pitch concept. Norma anticipated a specific purpose for this meeting; however, Tom was not aware of it or of what he might be asked to say. At this point, Tom did not know how this meeting would progress; later, he mentioned that Edwards and Dan did have elevator pitches and seemed to have more experience with them: "And [David Edwards] went through the whole thing, showing them a map and how it looked and the Corps and how the water level and all, and he throwed his elevator pitch." Dan had had some experience with organizing and speaking about conservation efforts while working with an oil company in Alaska, so Tom expressed confidence in the experience these speakers had. While Tom may have been frustrated with the lack of organization in advance of this meeting, he also lacks an idea of purpose for visiting the XYZ Company. In the conversation with Norma before the meeting, Norma's plan for reviewing the elevator pitch contrasted with Tom's lack of not only a plan but also a pitch. While Edwards obviously had a plan, Norma was able to contact Tom more easily beforehand to attempt to shape it. Even if Tom did have some idea of what to say to the XYZ Company, based on his involvement in past community meetings, denying a formal plan thwarted Norma's attempt to control the conversation remotely, and notably, he did not consult with Norma beforehand on how to develop an effective elevator pitch even though Norma had experience with this concept. Tom resisted Norma's regulating behavior by not participating in developing this pitch with her while allowing himself the opportunity to participate in the meeting as he saw fit at that time.

Pulling Up "Irrelevant" Information on the Internet to Try to Start an Organization

While Norma presented the competent ethos of an efficient organizer when she facilitated the first town hall meeting based on appeals to credibility and experience, later, community members' faith in her decreased when she did not seem knowledgeable of local regulations for starting a nonprofit organization. Because information on developing nonprofit organizations is available online, Norma began to research the process while also relying on her experience working with nonprofits in California. Despite her research, community members did not have confidence in the results. As Denise said, "She pulled it up on the internet, and she wanted to tell them how to do it [construct a nonprofit organization] off the internet. Well, you know, you can't run [name of current state] like you run California." Even though Norma had had some experience working with nonprofits and grant writing, that expertise did not transfer to this community work, in part because of her reliance on past experience in another state and conducting internet research that didn't seem locally relevant. In addition, community members also criticized the Convention of States organization because it was "something she pulled up off the internet," according to Denise, and no one in the community was familiar with that organization. When Norma and Tom disagreed about where Norma was getting her information on Corps' motivations on which to base her organization efforts, Tom retorted, "No, I don't know where you're getting this information from, but you can't believe everything you see on the internet." This distrust seemed to originate from Norma's access to information that the community was unfamiliar with and didn't have access to. Norma's access equaled control, and the community had no idea where the information was coming from and how it might or might not relate to local Grey Cliffs concerns.

Regulating Access to Information by Allowing Only a Select Few to View Her Grey Cliffs Facebook Page

At the beginning of the Grey Cliffs conflict, Norma created a Facebook page where she could post updates about meeting times, changing enforcement of regulations, and cleanup efforts. Norma managed this page and therefore controlled the information appearing there. Tom mentioned that her postings were negative and focused on shutting down the area to reduce crime. Also, Norma set the page to private so that not everyone could access it. "But the way her Facebook page was, was you had to answer three secret questions," Tom clarified. When I asked, "Like what kind of secret questions?" Denise responded, "We don't know. She didn't tell us all." Another interviewee, Felicia, thought that the main concern was security: "Well, didn't, wasn't it one of those boxes with the ABCs there, different ways you had to copy it to get in, for security?" Tom replied, "I don't know." In response to Norma's private Facebook page that regulated access and granted her management privileges, Tom created his own Facebook page that did not have such restrictions. "I don't want mine to be private; I want mine to be anybody that's on Facebook can go to it. And that's the way mine is," he stated. Norma's focus on privacy led to questions and suspicions among the community: why would anyone need to request permission to join this private group? Wasn't interest in this issue or community membership enough to justify roles as participants in the discussion? Rather than enable Norma's communication with this community, the private Facebook page became part of a regulating and distancing mechanism between Norma and the community, especially in contrast with Tom's page that did not focus so much on limiting access to information. This distance increasingly alienated Norma from the community and made establishing a relationship more difficult, as further evidenced by Norma's lack of personal connection with the community.

Lack of Personal Connection with the Community

Norma pursued credible ethos development, social action, and agency to accomplish change. Her motivations stemmed from personal interests and the desire to protect her property and its private access to the lake, as she had mentioned in her interview; restricted access would no doubt benefit the community and allow the area to rejuvenate environmentally as well. Norma put forth great effort to accomplish this social action by motivating the community to take part and act to protect the area. She composed the fliers, led the meetings, and began conversations with local and government officials. Yet distance existed and persisted between her and the community throughout the conflict. Some of this distance was created by Norma, and some was constructed by the community as, for example, they focused on the fact that she was from a different state.

Personal Anonymity and Distance

No one knew Norma's last name. Fliers she distributed to community members about meetings referred to her simply as "Norma." Norma included her cell phone number on these fliers and encouraged community members to call or text her with questions, information, and ideas, but her last name was absent. Complicating this was the fact that her Facebook name was "Suzie Escapes" on the page she had created to facilitate communication about Grey Cliffs. This pseudonym remained a topic of discussion as some community members elaborated on Norma's role in the conflict-management process. When I began asking questions about Norma's involvement, Tom didn't even refer to her as "Norma" but as "Suzie Escapes": "I call her '[Suzie Escapes]' or whatever. I don't know what her last name is." "That's what she is on Facebook," Denise clarified. When Denise asked Norma about this pseudonym during one of the meetings, Norma told her that she prefers to remain anonymous during this work, insisting that she could not have her name on any of the grants or proposals the community produced, a statement that generated even more suspicion. "Why couldn't she have her name on these materials?" I asked. Denise replied, "I asked the same thing, girl!" Felicia elaborated, "I think she's been caught in some illegal dealings or something." Denise continued:

> I asked her, I said, uh, 'cause she was saying she couldn't have her name on nothing, I said, uh, is that why you go by "[Suzie Escapes]" on Facebook? She said, "well, uh, that's part of it," she said, "part of it I sell stuff on the internet," and she said, "I meet them places; I don't want them coming to

my house." And she got a gate up before you get to her house; you can't [get] back there to her house. And somebody else went out there one day, and went all the way to her house, and she wasn't home. She was probably there and didn't answer the door, but anyway.

Throughout the interviews, community members referred to "Suzie Escapes" when answering questions about who was responsible for negative communication and events surrounding the Grey Cliffs conflict. For example, someone interested in visiting the area mentioned to Tom that, based on Norma's Facebook page, he had heard the area had been closed already. Tom related to me, "Well lo and behold, guess who told him it was going to be closed? [Suzie Escapes], [Norma]." Whenever community members mentioned the pseudonym, they associated it with facilitating negative activity at Grey Cliffs, which to them meant shutting it down. Norma's use of this pseudonym online contributed to her distance from the community, emphasizing even more that someone who was not "legitimately" connected to this community was proposing changes to Grey Cliffs.

Being an Outsider From Another State

In this locale that privileged a common cultural history, shared narratives, stories, and counterstories about Corps involvement and its relationship to Grey Cliffs, and generations of families with connections to the area, the fact that Norma had lived for a long time in another state bothered the community members. Even though Norma had shared in her interview that her connection to Grey Cliffs included "liv[ing] in and us[ing] the area for 30+ years," she had spent time working in California, apparently doing work similar to what she was trying to do at Grey Cliffs. However, as Denise stated when Norma began helping the community determine what to do in response to the Grey Cliffs conflict, "Well, you know you can't run [name of state] like you run California." "We're an entirely different state," Denise and Felicia agreed simultaneously. This outsider view led the community to look upon Norma's internet research into grants and nonprofit organizations with continued suspicion.

Interestingly, Norma was aware of this outsider status; she reflected during her interview, "It has been very difficult to overcome the 'Hatfield & McCoy' Prejudice of the ol' timers, as well as the ol' timers' prejudices against the newcomers to the area." In her ethnography of white and black working-class communities in the South, Heath (1983) distinguishes between "oldtimers" and "newcomers" in a way that clarifies that relationship in the Grey Cliffs' community:

> White oldtimers have long-standing family ties which link them to nearby farmlands; local businesses—hardware stores, department stores, etc.; or legal, medical, clerical, or educational roles in the region. Some are from the old textile families who founded the mills earlier in the century and still carry strong political influence across the state. White newcomers have come to the region through relocation by Northern industries, and they now consider the South their long-term future home. It is the rare newcomer who becomes actively involved in leadership in city politics, churches, or even the schools, since oldtimers look on newcomers in these positions with considerable caution. (pp. 237–238)

While Heath focuses on differences between white and black communities and socialization, social change, education, and other values, her definitions of these two white working-class groups in the South describe the Grey Cliffs community well to some extent: while Norma was reputed to be from California, rather than the North, she was not to be trusted and was to be looked upon "with considerable caution." She did not have a history with the area, at least not one as longstanding and significant as others who had the generational, cultural memory of the Corps and its relationship to the creation of the lake and the management of Grey Cliffs. Rather than these two groups being fixed, however, their dynamics can change: "'Insiderness' and 'outsiderness,' however, are not fixed or static positions; they are complex, fluid, and ever-changing" (Dunn, 2019, p. 96). This fluidity helps explain why Norma, even though she was considered an outsider and no one knew who she was, could begin a strong organizational effort and recruit the community, mostly consisting of insiders, to participate. Norma's potential to negotiate ethos was fluid as well, based on her efforts to negotiate a leadership role with this community.

To add to the lack of trust developing between the community and Norma, in part based on her outsider status and potentially differing values, rumors circulated that Norma had moved from California to escape legal troubles she had encountered during her nonprofit work: "She didn't spend [grant funding] like it was supposed to be spent or something, but that's why she left California and she moved here," Denise clarified. These rumors and suspicions damaged Norma's ethos in the eyes of the community. Not only was she an outsider who moved to the area because of an ethical failure; she was trying to anonymously regulate how the community organized efforts in response to this conflict with the Corps, by avoiding the use of her real name on any legal documents related to creating a nonprofit organization benefitting Grey Cliffs. In addition, she facilitated distance between herself and the community by remaining anonymous digitally via social media while also remaining physically isolated at her home, demonstrating some degree of physical separation as well. In fact, "Her house is really in [an adjoining] County," Denise offered, a fact that generated even more suspicion about Norma's involvement and motivation, since her residence wasn't even within the county experiencing the conflict. The community's discontent was perpetuated further with a growing sense that Norma was not to be trusted for other reasons.

Untrustworthiness

Over a short period of time, Norma's ethos of credibility and experience deteriorated based on rumors and poor reputation. Three additional themes supported the idea that Norma could not be trusted: her suspected support of the Corps' closure plan and related Facebook posts promoting erroneous and inaccurate information about Grey Cliffs' status, the rumor that she had been involved in illegal actions and unauthorized Corps activities, and her observed disloyalty and "stabbing people in the back."

Support of the Corps' Closure Plan and Related Facebook Posts Promoting Erroneous and Inaccurate Information About Grey Cliffs' Status

Although I did not hear Norma state publicly or in her interview that she wanted to close Grey Cliffs, several community members mentioned to me during their interviews that she wanted to close the area. Tom recalled:

> A lot of [community members] thought she was trying to close it. I have no idea, you know, just to have it more private for her. She's got a trail that goes down right down [to the lake], you know, 'cause she was telling us about her swimming and everything down there when everybody leaves out.

Tom later stated, "I think that's what she was trying to do [close the area]. And if you ask a lot of other people, they really think she had her way to it to close it down that she was really involved in doing that." In agreement, Denise stated that during one of the meetings, she had approached Norma and one of her neighbors and said to the neighbor, "I heard you and [Norma] was two trying to close it down." At the time, everyone at the table agreed. Lee, a community member whose land bordered the Corps land, verified,

> [Norma], I think her ultimate goal was to shut it down. She, she really, she would say in some ways she would say that she wanted it cleaned up and get the crime out of there, but, but she was also, you know, away from everybody, she made comments, of, we just need to shut it down, shut it

down, shut it down. I think her ultimate goal was to shut it down. Just hearing her, you know, speaking to individuals and so on and speaking to me, she said it to me, so. Um, I think she really was leaning more towards either shutting it all down or having just only community access, you know, not, not, you know, visitors and so on, you know. So . . . I think she wanted to put a tight rein on the area.

However, Lee, along with others whom he characterized as "the older guys, the older gentlemen that, that grew up in this area" wanted to keep the area open and accessible for the broader community and other visitors to use.

Interestingly, the Corps did not set out to close the area at first; that was just one of the options. As Edwards emphatically stated about the first meeting, "I thought I was pretty open that I don't want to close the area, I don't want to close the area, I said that five or six times." Yet, based on testimony from community members, Norma wanted the area closed. Norma appeared to have a personal interest in closing the area: crime would be reduced or eliminated if public use was prohibited, she would have more private access to her swimming area, and her private property would be left alone. Nothing was wrong with these motivations, but they stood in contrast to most community members' wishes to keep the area open and available to all. Tom stated that Norma had even started promoting on her Facebook page that the area had already closed, even though it had not; people had accessed Norma's page and then messaged Tom, seeking clarification about the closure they had read about on Norma's page. Tom responded with a correction to these people:

> It's not closed; you're just going to be able to camp on one side until they get another side cleaned up. And I said, one side is going to be camping; the other side's going to be for boats and parking, and I said we're trying to organize it to where it's a better place for everybody. But she has put a lot of, she tries to put a lot of negative, she don't no more now much here lately since everything's kind of going not her way.

Furthermore, Tom indicated that when Norma began posting erroneous information and comments on *his* Facebook page related to Grey Cliffs, he deleted all of her comments:

> See you know, I've got the, the Facebook page, [Grey Cliffs], Supporters of [Grey Cliffs], I got. And there's a lot of comments that people's made comments, you know, a few of them's like oh, they're shutting the trails down; I can't believe that. Which [Suzie] made the comments on there, and I'm sorry, but I deleted every comment she's made. I delete them all because they're so negative.

According to Tom and Denise, part of Norma's motivation to present negative and inaccurate information on Facebook could have been related to the idea that the Corps had only recently taken control of a bluff adjoining Norma's land that led down to the lake. Norma thought that she owned that land and wanted more control over it, but when Tom consulted Edwards about that, Edwards said, "We've [the Corps] always owned that; she just thought she owned that." Regardless of Norma's motivations for wanting the area closed, these rumors and suspicions fueled community distrust in Norma and her efforts.

Being Involved in Illegal Actions and Unauthorized Corps Activities

Norma had stated to several community members that she had experience with nonprofit organizations and grant work in California, but she could not have her name on any documents related to Grey Cliffs and that work in the current state, due to previous, suspected illegal activity surrounding grant work. As Denise mentioned, "So she couldn't have her name on none of the grants or nothing here." Felicia attempted to clarify, "The money or bookkeeping or anything?" Denise verified that no, Norma could not have her name on any of the documents but wanted to direct the process of forming the nonprofit organization and efforts to obtain grants based on her past experience. This process of prescribing recommendations for creating a nonprofit was interpreted by the community as "telling us what to do," yet, because Norma did not have any experience with the current state's regulations, she conducted online research, only to have community members, such as Denise, say, "She pulled it up on the internet, and she wanted to tell them how to do it off the internet." The community expressed skepticism as a result of these research efforts because, as Tom stated, "You can't believe everything you see on the internet."

Another issue with unauthorized activity involved rumors that Norma actually wanted to continue using the Grey Cliffs area for target practice, even though that was one of the activities prohibited by Corps regulations that had caused some of the environmental degradation of the area. Denise mentioned, "Well, uh, [Norma] though admitted that she liked to go down there and target practice." Tom added:

> Yeah, she wanted to know why there couldn't be a target range. I said, "[Norma], it's Corps property; you know, you can't go down there and shoot on Corps property without permission." And I said, "There can't be a shooting range down there." Well [what] if you put up a target? There again, you're putting up a target on a tree that's a Corps tree; you're shooting against something, you're tearing up, so you're destroying property, I

said, so. That's not going to happen; that's never going to happen. And I said something to [David Edwards] about it, and he was like, "you can tell her that ain't never going to happen."

If no one else could target practice without risking prosecution, Tom felt Norma shouldn't either.

The community members, based on these interviews, held differing perceptions about illegal and unauthorized activities depending on who was committing them and how the community was informed about them. From the very beginning, Edwards had stated that one reason he had to propose "severe" action, including possible closure, was because over the past 2 years he had received the 90 pages of documented dispatch reports from the local sheriff's office indicating all types of crime occurring at Grey Cliffs. The community rejected this documentation because members had never seen the crime occurring or evidence for it.

However, just suspicions of Norma's illegal involvement in activities with her work in California and her desire to continue target practicing on Corps land prompted distrust from the community. While Edwards possessed the dispatch reports as documentation, no official documentation existed about Norma's illegal actions or the rumors that she wanted to continue her target practice—at least, not to anyone's knowledge. Only her words could be used as evidence. Yet, these words, in conjunction with other parts of Norma's communicative ethos such as her supposed support of the Corps plan to close the area and negative, inaccurate Facebook communication, created a persona that the community simply could not trust. To the community, these efforts stemmed from her ultimate desire to have the Grey Cliffs area closed to the public, and the community had rejected this goal.

Disloyalty and "Stabbing People in the Back"

When Tom began suspecting that Norma actually was trying to close the area and was working, perhaps behind the scenes, to accomplish this action with local officials by controlling communication and posting erroneous information on her Facebook page and Tom's, he took this communication as a personal affront:

> But it, yeah, [my creating the Facebook page in opposition to Norma's] was kind of done intentionally, too, like I say, you know, against [Norma], so. That was my intentions, you know. Not that I hate her, I don't hate her or nothing; it's just, I don't like for nobody to lie to me, to my face. And then turn around and stab me in the back, you know, like. I think that's what she was trying to do. And if you ask a lot of other people, they really

think she had her way to it to close it down that she was really involved in doing that.

Because Norma's Facebook page promoted inaccurate information about Grey Cliffs and what was allowed and not allowed, Tom concluded that the rumors and suspicions were correct: although Norma was trying to organize the community to act, her intention was to shut the area down or enable the community to take ownership of it in the form of a nonprofit organization that could regulate access to the area even more to a chosen few. This goal contradicted Tom's goal of keeping the area open for community access, and he considered Norma's misinformation on her Facebook page as lying to him and the community. In contrast, Tom's Facebook page reached out to the community with updated information about regulations in hopes to continue to draw people to the area and generate interest for future fundraising activities. Norma's continued efforts to facilitate either closing the area or creating the nonprofit organization to manage it was an act of "stabbing Tom in the back" by completely going against Tom's goals, which Norma was familiar with due to the collaborative involvement in the community town hall meetings. This bold opposition from an outsider, even though it was documented by somewhat shaky evidence, received very little tolerance from this close-knit community.

Rumors circulating throughout the community about Norma, as well as her developing reputation, harmed the ethos Norma appeared to negotiate at the onset of her role as community organizer. Despite her previous work experience, longtime connection (although seemingly somewhat tangential) with the area, organizational skills, authoritative demeanor, and skills researching and communicating with governmental officials, the community ultimately held on to the rumors and suspicions that Norma was not actually trying to act on behalf of Grey Cliffs and the community's desires. While early on Norma stated her strong belief in organization and logical communication and that she "[did] not think with [her] limbic system" as someone who considered herself "a scientist," she did attempt to communicate her values to the community, presumably as an effort to reach out and connect. She demonstrated these values by presenting how she thought communication should take place about solving Grey Cliffs' problems and reinforcing these values through texts. These values served to develop Norma's persona as she attempted to connect with the community through a perceived appeal to common values. As a community organizer, Norma attempted to regulate action through these values and texts, discussed next.

ORGANIZING TEXTS

Despite Norma's focus on face-to-face communication of values within this small community, she also made use of several types of texts that she implemented during her role as community organizer, to regulate and motivate the action of the community during this conflict. These texts more specifically reflected Norma's values as an organizer: Convention of States literature, fliers advertising meetings, agenda and meeting notes, and nonprofit organization rules.

Convention of States Literature

At the first meeting, Convention of States literature was prominently displayed in the form of a small brochure distributed at tables. The text "Washington is Broken" appeared at the top, followed by "But We Have a Solution as Big as the Problem." Above an American flag graphic that lists state names and "USA" where stars would normally be, the following text appeared:

> Article V of the Constitution allows the states to call a Convention of States to propose constitutional amendments to limit federal spending, debt, and regulations. Amendments can also create better checks on the judiciary and restore the Constitution-as-written rather than the Constitution-as-interpreted by activist judges.

A similar statement is included on the homepage of the Convention of States website (https://conventionofstates.com). While this literature was not explicitly discussed at the meeting, it did elicit two different types of responses from meeting participants related to values this brochure conveyed. First, in my interview with David Edwards, seeing this information indicated to him the possibility of "anti-government sentiment" that could be displayed at the first meeting: "So one of the individuals that uh helped set the meeting up is a [member of the] Convention of States. And so, there was already this, I had a notion, I had an idea going into it that there would be some, um, potential for anti-government [sentiment]." As a result of this awareness, Edwards decided to construct his presentation in more of an open discussion format.

Second, when I questioned Denise about this organization, she indicated she did not know what the organization was about and said dismissively, "It's something she pulled up off the internet." Tom agreed, "She pulled, yeah, a lot of it she would pull up off the internet. She'd tell us." Not knowing what this organization was about, Denise began connecting it to other seemingly inapplicable information Norma had found online about beginning nonprofit organizations: "She'd tell you how to go about getting all these loans and how to do this and how to do that, but she knows she had to leave the state of California to come here, uh, so we don't want to be in that boat." Norma knew that this area was conservative politically and perhaps thought that the Convention of States literature would resonate well with other conservative community members who would want to make a change. The presence of this information at the meeting did not persuade the community to act, though; as Tom said about the literature, for example, "I never did look it [the organization] up." Rather than connecting and identifying with the values this brochure promoted, community members viewed it with apathy and didn't express interest in even reading it.

Fliers Advertising Meetings

Norma created meeting fliers to motivate the community to attend the town hall meetings; these were handed out to neighbors by Norma personally as she canvassed the neighborhood door-to-door, and they were posted at the general store where the meetings were held. Below is a reconstructed version of one of the fliers with identifying information redacted. This flier advertised the second town hall meeting.

FLIER ADVERTISING SECOND COMMUNITY TOWN HALL MEETING
COMMUNITY MEETING ABOUT [GREY CLIFFS]

 *Tuesday   *October 16   *6:00 PM
 *General Store   *123 Main St.

 This will be a WORKING MEETING of anyone interested in finding feasible Solutions to Address Problems at, and Achieving Goals of the Proposed Closure of, the [Grey Cliffs] lake access . . . & unintended consequences impacting the Community.

 Neighbors are strongly urged to attend. Positive input is acceptable—*Please leave pitchforks & negative attitudes at home.

 Government Officials are invited to come Listen & Advise the Neighbors. This is a meeting for Grassroots Input; not a meeting for "Govt Presentations."

 Contact [Norma] for more info. . . . Phone or text 555-555-5555

 Request to join Facebook Group "[Name of county and state] Neighbors for [Grey Cliffs]" to submit ideas in absentia. [URL for Facebook group]

Through this flier, Norma addressed the possibility that the community might not be open to government presentations, given what the community had already heard from Edwards and the community's resistance to his ideas. Anticipating this resistance, she encourages the community to "leave pitchforks & negative attitudes at home." The focus is on generating solutions from the community, as well as the officials, who "are invited to come Listen & Advise the Neighbors." Possibly addressing the audience of government officials as well as the community, Norma stresses that "this is a meeting for Grassroots input; not a meeting for 'Govt Presentations.'" The "Grassroots input" also refers to the strong sentiment in the area to avoid federal regulation; the community, based on Norma's plan, would necessarily want to take this area over and control it at the state or county level, rather than at the federal level. While this control is implied through the reference to "Grassroots input," Norma also includes strong wording in the flier's first paragraph that had the potential to resonate negatively with the community: "problems," "proposed closure," and "unintended consequences." Presenting this context so clearly at the beginning of the flier set the stage for the focus on "solutions" and "goals" at the upcoming meeting as well as opened the door for further communication with Norma via phone and the Facebook group she established. Despite Norma's firm rejection of emotion in communication, the flier she created contained some strong, emotion-laden language that did resonate with community members initially and motivated them to attend the meetings. These brief attempts to construct emotional appeals that resonated with the audience did, therefore, work.

Agenda and Meeting Notes

Norma created an agenda for the second meeting and also typed up notes from the first meeting (see Appendix 5.A and Appendix 5.B). These documents were designed to guide the discussion. The agenda included goals, problems now, problems caused, solutions, feasibility of ideas, and a plan of action going forward regarding discussion. This organization reflected a similar pattern that the first meeting followed, based on the notes/minutes Norma typed and the flip chart notes she took during the first meeting. In these minutes, the long-term goal was listed as "shift responsibility for some of maintenance of Goals to Outside Public or Private entity (Out-Grant &/or Sub-Lease)." Norma also included a lot of specific solutions discussed at the meeting, such as "turn area into a 'County Park,'" or "turn area into a 'State Park.'" The agenda regulated continuity of the discussion themes between the first and second meetings. The meeting notes/minutes also provided continuity in that people not at the first meeting could read what had been discussed there and ensured that the focus remained on the long-term goal of shifting responsibility from the Corps to another responsible entity.

Researched Nonprofit Organization Rules

One characteristic about Norma's reputation that kept circulating among community members was that she had pulled up irrelevant information from the internet to help start a nonprofit organization as a solution to dealing with the Grey Cliffs conflict; this information was also associated with the rumor that Norma had been involved in illegal activity in California and was attempting to localize her previous expertise and work experience to this community. The rules Norma found also connected with her value of consistency: determining what rules should be followed corresponded with developing a clear organizational structure. Only by following a clear organizational structure would the community have any chance of creating a nonprofit organization, and, according to Norma, creating a nonprofit organization was the solution to solving Grey Cliffs' problems. When these guidelines were distributed, the community had developed several subcommittees to address the issues at Grey Cliffs, including generating material that could be used in developing a nonprofit organization, as well as recommendations for creating the various committees.

The general organizational guidelines Norma found are presented in Appendix 5.C; not only does the document present the information in numbered steps with headings, but the guidelines also included personalized comments from Norma, such as "take plenty of paper notes, but make notes in the 'comments' of this post, also." This statement indicates the multimedia format in which the guidelines were presented: they were given in hard copy form at the meeting, but Norma had also posted them on the Grey Cliffs' Facebook page. Such an electronic format would allow committee members to type comments while meetings occurred, resulting in an archive for the meeting notes but also allowing for Norma's continued supervision and regulation of activity based on the comments posted. Each committee category contained a list of goals and tasks. At the end of the document, Norma wrote, "These ideas are by no means 'The Way' to organize our group. These are not carved-in-stone Committees. Please develop the ideas that will accomplish the Corps Goals for [Grey Cliffs]." After all, Norma attempted to convey her belief in "Grassroots efforts" and wanted to recruit community involvement to accomplish the goals she had listed for each committee. Fulfilling these goals and tasks on her own would have been impossible; Norma wanted the community to invest in and take ownership of these ideas that she began. Norma's communicated willingness to collaborate with the community and move it forward motivated the initial community attendance at the early meetings and generated a sense of hope that, through working together, something could be done to resolve this conflict.

All of these organizing texts—the Convention of States literature, fliers, agenda and meeting notes, and organizational guidelines—provided guidance and motivation for this community to negotiate change with the U.S. Army Corps of Engineers using organized, clear-cut strategies while still providing room for community input, choice, and ownership of the negotiation process. Each text also coordinated with Norma's values of face-to-face communication, focusing on facts, and lack of reliance on emotion. Based on the Convention of States literature, Norma encouraged this community to participate in grassroots efforts and provided an organizational framework to prompt this community to start their work. However, from the very beginning, the rumors and reputations building around Norma's ethos deconstructed Norma's overall persona, resulting in reduced agency.

CONTINUED COMMUNITY RESISTANCE, CONSTRAINED AGENCY, AND LACK OF ETHOS NEGOTIATION

When Norma assumed the role of community organizer in helping solve Grey Cliffs' problems, she appeared to have a strong "resume" listing her experience and qualifications that made her very well suited for this role. Her logos and appeals to credibility and experience were strong; however, appeals to ethos through character, such as sincerity and affinity, were lacking in her community organizer persona. As George Campbell states, "All the ends of speaking are reducible to four; every speech being intended to enlighten the understanding, to please the imagination, to move the passions, or to influence the will" (Campbell, 1776/1990, pp. 749–750). While Norma "enlighten[ed] the understanding" of the community with her knowledge of the situation and her skills in organizing, her communication could have been even more persuasive if it had also accomplished the last three aims of discourse Campbell mentioned: "to please the imagination, to move the passions, [and] to influence the will." Not only did the community resist Norma and her efforts to form a nonprofit organization, but from the beginning, it also worked against her to deconstruct her agency and, ultimately, her ethos, through antenarrative and counterstory. Her rhetorical persona was incomplete, and she was not able to accomplish the action she desired due to this constrained agency.

Constrained Agency

At the beginning of the Grey Cliffs conflict, Norma felt that her efforts at organizing the community were working. Negotiating agency with the community had been successful "b/c my ideas were logical and fact-based," and "such a large turn-out for public forums" indicated community interest and support. However, shortly after the first meeting that Norma led, she herself realized her agency had been constrained and basically deconstructed by the community. While Norma believed that "there are rules and formalities established and followed within any organization. You can't just show up and do what you want, when you want to, etc.," in contrast, the community did not accept the rules and recommendations that she presented (see Appendix 5.C). Norma reflects:

> Ppl who volunteered for committees did not follow my written guidelines, nor were they familiar enough with "business formation" to see the importance of the "process." Some committee volunteers misconstrued volunteering for actual appointment to an office—the committees were [supposed to] set up the process for electing or appointing a board of directors and other officers "by the membership," ways to interact with a "membership," and with public officials.

One of Norma's clear values was a reliance on a set organizational structure and following the rules; after all, her goal was to create a nonprofit organization that would be able to enter into a legal agreement with the Corps to govern the area. However, what she didn't expect was that the community might be happy with a "compromise," that a positive relationship could be negotiated between the Corps and the community, and that an alternative solution could be developed between the two groups that satisfied the community, at least for the time being, leaving Norma the "middle woman" out of the equation. Norma reflected further:

> I was left out of all further meetings of the volunteers, and what was said in those meetings, b/c of their misconceptions. (I did make personal phone calls to many of the committee members when I heard of happenings of a crucial nature.) Therefore, nothing moved forward. When I and another woman, who'd said we'd do "research," made inquiries with accountants, attorneys, business assistants, etc., about forming a 501(c)(3), they uniformly responded that we were not ready—b/c none of my written objectives had been met (the guidelines published on Facebook to be used at the first committee volunteers meeting). Being a "volunteer" organization, demands may not be made of the volunteers. All that can be done is to find ppl who understand what is required—or are willing to learn—to become a functioning "organization," to set goals, and execute plans. Part of the function of the committees was to FIND THESE PPL, but, as I said, the committee ppl thought they were "it," and not understanding the

process, nor being willing to learn, decided that getting the CoE [Corps of Engineers] to "do something and keep Grey Cliffs open" was all that was required. Done. So, back to square one.

As a result of this constrained agency, at least during this time, Norma discontinued her community involvement, although she still indicates interest in these issues, which she considers unresolved. The community's disconnection and lack of relationship with Norma are examples of how an audience can deconstruct a speaker's ethos, based on lack of trust (Mackiewicz, 2010) as well.

Lack of Ethos Negotiation

Norma's ethos appeals to credibility and authority as a community organizer were impacted by key community members in several ways. A couple of members significantly involved in the Grey Cliffs conflict supported Norma's role in the beginning. Edwards mentioned during his interview that "[Norma] . . . did a good job on social media, weeding out the negative comments, you know, stay focused, focus on this, you know." He credited her with initially motivating the community. And Paul, a community business owner for whom closure would have a severe impact on the number of customers at his store, said,

> See, she realizes a lot of people don't have the internet, so she goes out on her own and puts the papers [fliers] in their mailbox. She's instrumental. No, she was instrumental in that, and she brings them [fliers] here [to the store], too, and I post it on the doors, you know. . . . Yeah, we can't just be on social media, like I was saying.

No one I spoke with disputed Norma's role in effectively organizing the community and motivating it to act in the beginning. However, as time went on, other community members indicated their distrust in Norma, as they voiced counterstories, many ignited by rumors. These counterstories fed into an antenarrative that deconstructed Norma's power; essentially, the community treated Norma as they had Edwards because they believed Norma was also trying to close down the Grey Cliffs area, and the community did not want to support that type of narrative about Grey Cliffs' future. Once again, as this community illustrated with Edwards, appeals to credibility and authority alone are sometimes not enough to persuade an audience to act.

Tom and Denise were two key members who voiced this distrust in Norma through counterstory. This suspicion began the first time Norma visited them to motivate them to attend the first meeting. Norma began the interaction with Tom and Denise by stating that she had taken care of Denise's parents long ago. According to Denise, "she actually come to our house, knocked on our door, and she didn't know who we were; and [Tom] went to the door, and then she come in, she was saying that she used to take care, she was a home health nurse, and she actually took care of my momma and daddy. Done therapy with them." Tom continued,

> Well, she kept asking me if the [Browns] lived there; that's what she [Norma] was—a [Brown], and I said, uh, "No." And she kept saying, well Ma and Pa used to live here. And I said, "You talking about [Bill] and [Sarah], my in laws?" Uh, yeah, and I said, "Well, they lived here; we've always owned the property and the place but they lived there." And I said, "My wife," she was setting there, I said, "that's her mom and dad."

Norma attempted to make this insider connection to let Tom and Denise know that she had a history with this community; this was a strategy to build her credibility and also affinity, a way to justify, to some degree, her interest in the Grey Cliffs conflict. However, after establishing this connection, the conversation turned to Grey Cliffs. Tom relates,

> And she was telling us all about [Grey Cliffs], you know, and I just said, "Listen," and I told her, I said, "I've been around here all my life," I said, you know, and I said, "She's [Denise] lived here." I said, "We know all about it [the conflict]," you know. And she was saying this was going on, that was going on, and I'm thinking, "Lady, that ain't been going on; if it has I've never seen it." Yeah, the crime, and all, I mean a lot of that I think has been made up.

Denise agreed, "It sure has." Tom and Denise believed that the crime narratives about Grey Cliffs were an excuse to close it down for other reasons—so that Norma could make the area more private for herself, for example.

This distrust in Norma's motives prompted Tom to exclude Norma from serving on the board that was created to help find solutions to the Grey Cliffs conflict. As he reflected on the second meeting when board members were ultimately chosen, Tom clarified what he thought Norma's role should *not* be, using her Facebook pseudonym:

> And I said, I really don't care if [Suzie] is here, I'd rather her not be because she's not going to be on the board, I mean, I just as soon told the lady, and she's like, why not? And I said, because we don't need her on the board. We need people that lives in this community, that's been in this community and wants to help it not hurt it.

Denise emphatically stated in support, "[Suzie] ain't running this meeting." Furthermore, Tom continued ignoring Norma's ideas by rejecting her recommendation to create an elevator pitch and run it by her before

Tom visited the XYZ Company with Edwards and Dan. Additionally, when Norma later continued promoting her idea of forming a nonprofit organization and supporting community takeover of the area, which were plans the community opposed, Tom objected:

> See I heard that she was wanting it more private for her, like if they would have shut it down, you know, and blocked it off, then you know, she would be able to ride her 4-wheeler down there and swim. You know, that was her big concern. She says, she kept telling me, you know, why don't the community lease it, why don't the community do this? Because it's going to cost a lot for the community to lease it, for us to lease it because, you know, because I talked to [Edwards] about it, he said, look, he said, you take it over and lease it, he said, you got to put lights up, you got to have running water, you got to, you know, be able to make it like. You got to have concrete boat ramps, and that takes a lot of money.

Tom rejected Norma's idea to create the nonprofit organization and instigate community ownership of the area because he didn't trust Norma, was suspicious of her motives, and knew the community did not have the funds to create the physical infrastructure needed to keep the area open based on her plans and feedback he had received from Edwards.

At the beginning of her interaction with Tom and Denise, two influential members of this community, Norma had specifically tried to make a connection with them by stating she had taken care of Denise's elderly parents and that her parents were from the area; she attempted to connect with them by stating specific names, even, to indicate her knowledge of the family history of the area. In this sense, Norma attempted to appeal to Tom and Denise through a commonly held value of family and local connections in an effort to establish trust. However, Tom and Denise did not know Norma personally, their interactions with her were limited, and they didn't really have any information to verify what she had said. While approximately Tom and Denise's age or slightly older, Norma had moved across the country and had returned only relatively recently. This fact seemed significant to Tom and Denise, who did not accept Norma as an authentic community member. Somewhat unexpectedly, Tom voiced his opinion that, ultimately, Corps ownership of Grey Cliffs should continue, despite the community's conservative political climate that seemed ripe for a Convention of States takeover. The rumors surrounding Norma and her negative reputation fueled Tom's rejection of Norma, or "Suzie," as he often called her. While some meeting participants, such as Edwards and Paul, did respect Norma's work as it related to logos and her organizational work, her ethos, as Norma attempted to construct it, did not withstand the community's scrutiny overall, including the counterstories the community generated about her. Her appeals to affinity were not long-lasting or widespread. Her emphasis on logic overwhelmed the minor attempts to reach key target audience members through character development; her framing of solutions did not resonate with the values of this community. As a result, the rhetorical appeals Norma attempted to exert failed.

Lee, another community member, characterized Norma as "pushy" and attributed the community's rejection of her to her lack of attention to gender dynamics of the community, as well as her lack of being an insider: "This older group of men on the board didn't react very well to a woman coming in and telling them what to do," he said. "She seemed very pushy. And it was my understanding that she was from California. The community didn't react well to someone from the outside coming to the meetings and acting the way she did." Notably, Edwards and Paul did appear to value Norma's organizing efforts. Edwards thought enough of Norma's interest in the conflict to discuss options with her and supported her role as community organizer, a collaborator with him, in essence. And Paul had praised Norma's efforts to connect with the community face-to-face through meeting with neighbors and distributing fliers. As a woman, even as an outsider, Norma's stories of her own work experience and knowledge of the area were valuable. In her leadership role, she drew upon the authority and expertise that she had, but these experiences were not valued as knowledge within this community. As Petersen (2018) discusses in her work on female practitioners, technical communication, and social justice, oppressed voices can at times be marginalized when attempting to make needed changes within communities; these voices need to be heard, though, since they often represent other marginalized voices that may not ever be heard (p. 5).

The community's rejection of Norma was unfortunate, since it bears some similarity to other grassroots, environmental efforts begun by female leaders. Certainly, "environmental sexism" (Gaard, 2018; MacGregor, 2021, p. 237) plays a part in decision making about environmental issues. This marginalization exemplifies itself as scholars note that more women than men lead grassroots efforts because women, as a marginalized population, tend to work "outside the system" (MacGregor, 2021, p. 239), yet leadership positions in policy development and other environmental leadership roles tend to be held by men (Gaard, 2018), and many times white men (MacGregor, 2021, p. 242; Taylor, 2014). Women often lead grassroots efforts because of their social and cultural roles that include caring for others, rather than "an innate sensitivity to the natural world" (MacGregor, 2021, p. 239); they also have experienced marginalization and injustice themselves and are thus well positioned to understand the seriousness of environmental injustice (MacGregor, 2021, pp. 238–239). Since Norma worked as a caregiver in her role as a home health nurse, her infusion of care into Grey Cliffs' environmental concerns makes sense. Indeed, as MacGregor (2021) points out, "As those most responsible for providing everyday sustenance and taking care of children, women are typically more in tune with the quality of the local environment than men" (pp. 239–240). Based on this case study, Norma demonstrated confidence in her values, stemming from her care and concern. She was more than willing to act, and her efforts are admirable, even though they did not result in the types of changes she desired.

While not formally trained in the use of rhetorical strategies, Norma very successfully motivated the community to come together to begin addressing the problems occurring at Grey Cliffs. However, with a damaged ethos from the rumors and reputations the community assigned her and her tangential positionality with the community, Norma's knowledge of organizational strategies became less relevant in community members' eyes. In addition, her lack of emotional awareness and need to establish trust distanced Norma even more from a community that expressed such strong feelings and values about the area. While Norma negotiated agency well with the community in the beginning (or at least negotiated a position for that to occur), the community actively constrained her agency as time went on, severely limiting her power to act.

Both Edwards and Norma experienced community rejection based on ethos appeals grounded in authority, expertise, and credibility alone. This rejection was not as longstanding for Edwards, however, as it appeared to be for Norma. Once Edwards recognized the community's hostility and rejection of his initial message, he began adjusting his narrative and framing of the rhetorical situation to incorporate increased community connection through co-constructing an ethos based on his character, including appeals to affinity and sincerity. These appeals resulted in a growing relationship with the community. I present an analysis and discussion of this process in Chapter 6.

KEY RECOMMENDATIONS FOR TECHNICAL, PROFESSIONAL, AND ORGANIZATIONAL COMMUNICATION AUDIENCES

- Identify potentially marginalized voices and listen to them; encourage their participation in conflict resolution.

- Determine ways to incorporate gender balance when communicating, generating ideas, and creating documents for a variety of media technologies.
- Incorporate the input of marginalized voices into documents created to address conflict involving technical communication, such as grant proposals.
- Acknowledge and incorporate the experience of community members into document creation and communication opportunities, when possible.
- Encourage leaders and other communicators with strong foundations in logic and science to connect with influential audience members through meaningful relationships, based on affinity and sincerity, for example.

KEY RECOMMENDATIONS FOR ENVIRONMENTAL SCIENCE AND PUBLIC POLICY COMMUNICATION AUDIENCES

- Identify key knowledgeable, potentially marginalized participants who could contribute experiential knowledge to developing public policy for localized environmental problems.
- Pay attention to balancing gender participation in developing solutions to environmental problems and creating public policies.
- Encourage participants to align with values and goals that contribute to the common good for other individuals as well as the environment.
- Help identify alternative solutions to policy conflicts that seem more feasible for local issues.
- Find ways to include dominant as well as marginalized populations when creating and modifying policies that impact public spaces.

APPENDIX 5.A
Agenda for Second Community Town Hall Meeting

Greeting & introductions
    Pledge of Allegiance
    Opening Remarks & "ground rules"

1. "Goals" of the Corps' Closure, as you understand them.
2. "Problems Now" at [Grey Cliffs]
    2.a. "Problems caused" by Closure.
3. "Solutions" you'd like to try.
4. Feasibility of ideas [will depend on]
    a. If the idea(s) meet the Corps of Engineers' Goals of Closure.
    b. Estimated Costs to implement the idea(s), & (1) to what entity.

5. Which are the controlling entities, & order of importance?

6. Choose one or more Community Representative(s) to take our ideas, without prejudice for their own ideas, to discuss them with Law Enforcement, County, State, or Federal elected officials or departments.

APPENDIX 5.B
Results and Notes for Agenda for Second Meeting, 10/16

Greeting & introductions

    Guests present: [David Edwards] Corps of Engineers, Sheriff [Wheeler], [Albert Jones] staffer of Representative [Tammy Phillips], [Jim Theriot] [name of county] Chamber of Commerce.

Pledge of Allegiance

Opening Remarks & "ground rules"

    This Meeting turns the order from the first meeting "on its head." Our Guests are here to listen and advise Us neighbors as we discuss Goals, Problems, and Solutions for the Proposed Closure of [Grey Cliffs].

"Goals" of the Corps' Closure, as you understand them.
- Save $$
- Save Habitat (shoreline, Corps Property)
- Stop Illegal Activity

. . . .

Long-term Goal:

    shift responsibility for some of maintenance of Goals to Outside Public or Private entity. (Out-grant &/or Sub-Lease)

2. "Problems Now" at [Grey Cliffs]
   - Not Enough Sheriff Patrols.
   - Illegal activities ie Theft, Burglary, Gunfire, Drug trafficking, Underage &/or Unruly "Partying," Trash-dumping, Off-Road motor vehicles.
   - Poor Road Surface/Condition.
   - Primitive" Camping ie No Bathroom, No fresh water, No trash receptacles, No fire pits, Destroying Habitat.
   - No, or Destroyed, Signage with Regulations/Prohibitions clearly posted.
   - Spotty, Unreliable Cell Phone Service

2.a. "Problems caused" by Closure.
   - Law-abiding ppl will be shut out.
   - Criminals will feel "safe" when they sneak in on foot or by boat.
   - No other [name of lake] access within 12 miles.
   - Local businesses will be adversely affected.
   - Ppl turned away from [Grey Cliffs] access road may try to find other roads to the lake through private property.
   - Illegal activity on private property will increase.

Decreased property values.

Emergency vehicles, including Fire Trucks, will be unable to reach the area by land, and will further lengthen response time for 911 calls.

Virtually all law-abiding citizens who use the area AND criminal element will be permanently angered.

3. "Solutions" you'd like to try.

Community Volunteers for: Patrols, Trash pick-up, etc.

Turn area into a "County Park."

Turn area into a "State Park."

Law Enforcement conduct "Sting Operations."

Security cameras.

Security "people," ie, private entity.

Block sides of road way to prevent Off-Road Vehicles, but leave access and sufficient boat ramp/parking area open.

Replace and increase signage posting regulations/prohibitions.

Enforce Penalties for First Violations of Laws/Regulations. No More "verbal warning" or "written warning."

Set up scheduled "community volunteers" to do Drive-thru monitoring of activity at [Grey Cliffs], who will call Law Enforcement or other emergency services as needed.

Get cell phone service improved in all parts of [Grey Cliffs].

Open a Neighbors' 501c3 (not-for-profit Organization) to be able to contract/make agreements with Govt entities, and fund-raise to complete projects desired by the Neighbors in accordance with contracts/agreements with the Corps of Engineers.

Research the availability of Govt Grant $$ to pay for any of the Solution Ideas.

(Although the Corps of Engineers & Sheriff [Wheeler] we're present, the solutions were not fleshed-out fully enough to complete items 4 & 5 below.)

4. Feasibility of ideas will depend on
    A. If the idea(s) meet the Corps of Engineers' Goals of Closure.
    B. Estimated Costs to implement the idea(s), & (1) to what entity.

5. Which are the controlling entities, & order of importance?

6. Choose one or more Community Representative(s) to take our ideas, without prejudice for their own ideas, to discuss them with Law Enforcement, County, State, or Federal elected officials or departments. This item was addressed by the people present by them "signing up" as volunteers. Each person will be contacted in the next week for a phone "interview." The interview will ask the following things: (a) what is your realistic estimate of your time, duration, and physical commitment to solving the "Closure Problem"? (b) how do you want to be involved: Community Representative to speak to Govt entities? Pick up trash/maintain or improve Habitat/Shoreline? Drive-thru monitor? Bookkeeping/Accounting for 501c3? Develop ideas (as a part of a responsive group of volunteers) and record them in a form that is presentable to Govt entities? Research? Other? Since the weather kept some

ppl away from this meeting, if you want to Volunteer for a "committee," please CALL or TEXT Norma at 555-555-5555.

NO ONE WHO WANTS TO BE ACTIVELY INVOLVED WILL BE TURNED DOWN!!

NEW IDEAS ARE ALWAYS WELCOME!!

APPENDIX 5.C
Guidelines: First Meeting of "Committees" for [name of county and state] Neighbors for [Grey Cliffs] Group

(Suggestion: take plenty of paper notes, but make notes in the "comments" of this post, also.)

1. Develop a Mission Statement for the Group . . . (should answer the following questions):

    "Who are We"?
    "What are We organizing to achieve"?
    "How are We going to conduct ourselves"?
2. Brainstorm the purposes for each committee.

    *Decide on "tasks," "duties," "who will be responsible for tasks & duties," etc., & how these things will be done.
    *Make a primary goal for each committee (Short & long term, if appropriate).

Here are a few examples of Committees' Goals, Tasks, Duties:

Executive:
1. Goal: efficient communication of Neighbors' wishes, ideas, plans to appropriate entities for implementation.
2. Task/Duty: incorporate sub-Committee recommendations into final plans
3. T/D: develop plans into formal written product
4. T/D: accept/collate/disseminate information from/to membership.
5. T: maintain written records of attendance, contacts made, summary of discussion, etc.

Research:
1. Goal: provide other committees with information pertinent to their functions and goals, i.e., find out "How Things Work."
2. Task: obtain names and phone numbers/emails of Govt officials whom the Executive Committee may need to contact.
3. T: find out how to become a non-profit organization.

4. T: gather info on how Corps wants [Grey Cliffs] maintained, & relate info details to Trash Committee.
5. T: gather info on how the Group can lawfully assist the Sheriff in "watching" [Grey Cliffs]; find out how successful programs have worked in other areas.

Trash:
1. Goal: maintain [Grey Cliffs] "area" (from entrance by [Grey Cliffs] county road to include accessible Corps of Engineers property and shoreline).
2. Long Term Goal: Develop plan for maintenance of area per contract/agreement w Corps/[name of county] if our group becomes one of the contractors.
3. Task: organize trash pick-up

Security:
1. Short-term Goal: determine best way to monitor activities at [Grey Cliffs] and alert Law Enforcement.
2. Task: decide how to schedule neighbors to accomplish goal.
    *Develop an organizational "tree."
    *Develop a "phone tree" &/or a way to communicate which is acceptable to all. Hybrid "text-email-phone"—anything possible.

—These ideas are by no means "The Way" to organize our group. These are not carved-in-stone Committees. Please develop the ideas that will accomplish the Corps Goals for [Grey Cliffs]:

- Save $$ & off-load responsibilities for its maintenance.
- Eliminate Criminal Activity.
- Restore the Natural Environment.

AND OUR GOALS:

- Keep [Grey Cliffs] OPEN for our lawful activities.

6

A CORPS RESOURCE MANAGER'S RHETORIC OF RELATIONSHIP
Co-Constructing Ethos With a Community

> "When I'm able to make a connection with an individual, I'm able to gain more compliance. This is in direct conflict sometimes with rules and regulations because rules and regulations are black and white."
>
> —David Edwards, Corps of Engineers resource manager

Both Edwards and Norma attempted to regulate this community in different ways: Edwards, in his role as Corps resource manager, possessed the authority and credibility, based on experience and expertise, to enforce action, such as closing Grey Cliffs. The rules and regulations developed by the Corps (see Appendix 6.A) held legal weight regarding any activities taking place on its properties, and Edwards was responsible for those regulations; essentially, his role was regulating behavior. Norma also attempted to regulate the community based on her previous experience and expertise in the workplace and her desire to form a nonprofit organization that ultimately might have the potential for promoting community or perhaps county ownership of Grey Cliffs, giving her more rights over her property and the area surrounding it. However, the community initially rejected both of these regulatory efforts, instead responding, involuntarily at first, very emotionally to these efforts and indicating through their own personal narratives how much their experiences and values connected to this beloved area.

At the beginning of this conflict, Edwards framed the issues using crime statistics and environmental damage, essentially focusing on his ethos as credibility, based on authority, experience, and expertise. In contrast, the community attempted to counterframe the issues with its discussion of the area's virtues, using stories that communicated affective values. This process did not work very well; the frames were still very different and did not encourage resolution, based on the narratives, values, and stories both sides were communicating. These conflicting frames harmed both Edwards's and the community's ethos by highlighting how polarized and uncompromising this government organization and community appeared to be.

While Edwards did possess and acknowledge responsibility in the way he framed this conflict at the outset that resulted in such violent opposition from the community, he later also acknowledged that he thought he had failed in this opportunity, at first, to address this issue with the community in a way it would be well received. Edwards did not intentionally seek to alienate the community further with his initial proposal to close the area; however, as this work demonstrates, Edwards had to then pivot and begin reaching out to the audience, constructing a revised ethos for himself and extending a willingness to co-construct an ethos with the audience that would yield the potential for joint, positive, just, social action. Edwards had to accept that "public perception shapes the context in which professional [communicators] operate, and it is important that they understand perception as part of their ethos" (Tillery, 2006, p. 326). Edwards had to acknowledge the community's perception of him as a governmental Corps representative and take that into consideration when addressing this audience.

In order for the community to trust Edwards, they would need to learn more about his character, and this character would need to convey a symmetry between values and qualities necessary for eliciting trust in the audience. These values and qualities would also need to reflect similar values and qualities that the community valued. Up until this point, Corps and community values did not seem aligned enough to accomplish positive change. In this case, Edwards would need to put forth an effort to establish himself as someone worthy of the community's trust who could then extend an opportunity for value alignment.

In this chapter, I present Edwards's ethos-negotiation efforts as he developed and presented his *character* to the community, rather than focusing on his credibility alone. Specifically, this character highlights Edwards's sincerity and affinity as he adapted his communication approach midstream, when he realized his initial regulatory emphasis was not having the desired impact on the community. This chapter analyzes Edwards's role as motivator of community action based on *relationship*, including strategies for communicating Corps values and texts through changed rhetorical strategies, including a revised map, rules and regulations, email messages, and stories.

BEGINNING TO DEVELOP A RHETORIC OF RELATIONSHIP

A key to developing community support, Edwards quickly discovered, was creating relationships with those in the community he deemed as influential. This need to enlist the community's help affirms Herndl and Licona's assertion that "agency is contingent on a matrix of material and social conditions. It is diffuse and shifting" (2007, p. 138). While Edwards clearly had the governmental authority to act on his own without the community, he also did not want to alienate them further, since the community already had a complicated, cultural-historical relationship with the Corps. Edwards was also required to inform and involve the community before he took any formal action as a Corps representative due to the National Environmental Policy Act, and, ideally, the community would be supportive of his efforts and actually be involved in them. To work toward this balance, this negotiation process, rather than communicating an authoritative ethos of enacting change through strong-arming the community, Edwards changed his strategy to begin negotiating a position of change.

Some might wonder why Edwards did not start with a rhetoric of relationship rather than a rhetoric of regulation; or perhaps he could have started off with a combination of the two. He did appear to have some limited, previous insight into this unique audience ahead of time. The answer lies in Edwards's estimated construction of his audience in advance of the first town hall meeting, which was based on conversations with a few community members. Only after the hostile outcry from the unexpectedly large audience in attendance that evening did Edwards realize that his regulatory communication strategies, based on credibility alone, were not only not working but also alienating community members in this expanded audience even further from what Edwards's ultimate goals as resource manager were.

In beginning to move from a focus on the rhetoric of regulation and compliance to the rhetoric of relationship in an effort to negotiate joint action—and, ultimately, socially just action—with the community, Edwards began to self-regulate and change his own behavior by introducing personal narratives, specifically about sincerity and affinity as part of his character, to motivate the community to act and make changes in their behavior. Edwards illustrates this process by first integrating his own personal desires and interests into the conversation with the community; doing so allowed him to connect with the audience and establish a clear sense of ethos through character development. Yes, his role was authoritative as a Corps resource manager, but incorporating ethos appeals of sincerity and affinity allowed Edwards to empathize with the audience as "one of them," to present himself through specific communication strategies as someone who could sympathize and empathize with the community's interests and upcoming changes to be made.

Edwards recognized that regulation alone, without relationship, often doesn't work; making that personal connection with individuals is essential. Edwards made an effort to connect with individuals in part because he knew this community's history with the Corps and wanted to maintain a positive relationship with the community. Knowing this context provided Edwards with additional cultural knowledge about his audience and allowed him to partially understand the community's initial perspective of hostility toward him; this type of resistance can often occur before rapport is established between community members and regulatory agency representatives. Community members may not fully understand what the representative's role is and may view the representative with suspicion and skepticism as a result (Williams & James, 2009, p. 92). In response to this hostility and uncertainty, emphasizing the character development of his ethos turned out to be a critical strategy for Edwards as he attempted to relate to this audience during this difficult conflict. As Tillery (2006) clarifies, "In addition to emotional appeals, experts also need to understand the role of the author's character in convincing an audience" (p. 326); through character, audiences can more readily understand the rhetor's goodwill, wisdom, and virtue.

Rhetorical Analysis of Character Development

To illustrate the complex, rhetorical processes at work as Edwards addressed and attempted to establish a relationship with this community audience in order to convey a character that the audience would accept, I analyze Edwards's interview transcript based specifically on appeals to sincerity and affinity, as they relate to ethos development. While many virtues can impact the ethos and character of a communicator, I focus on sincerity and affinity because these virtues stood out the most in Edwards's reflective self-narrative (the transcript of his interview). Table 6.1 shows this part of the analytical framework used in the study. While the Grey Cliffs community members initially rejected Edwards's ethos based on credibility, his character development did seem to reach this audience eventually, as this part of his ethos construction related more to establishing a relationship with the community.

Appeal to Sincerity

Sincerity for Edwards was linked to values such as being ethical and trustworthy, which are in turn related to goodwill. All of these values

Table 6.1. Rhetorical Framework for Analyzing Edwards's Narrative—Character

Ethos appeal	Explanation	Example quotations
Character		
Sincerity	Statements containing the word "sincerity" or others conveying honesty and truthfulness.	"How I'm treating them, how I'm providing information and relaying that information in a positive manner help take the edge off of the message that they're receiving. And maybe opens their mind, broadens their perspective a little, and then helps me form more of a connection when they see that I'm trustworthy and that I'm sincere in my desire to make a difference." "And I think that's what this all boils down to, in that, you know, I think they see, I hope they see, the sincere and willing participant that, um, I could bring a lot of resources to the table, and I'm willing to commit those resources to bettering the services that the public has at [Grey Cliffs]."
Affinity	Statements indicating similarity, commonality, friendship, or desired goodwill, especially relating to ideally aligned values.	
Expressing common interests		"I like to hunt, too." "I traveled all over the logging roads; I grew up on the strip jobs, and I loved off roading." "I like to fish, too."
Reaching out as a friend		"I just basically approached them and asked for their help. I said, 'Guys, we've got to stop this; this is something that we have to do; I need your help.' And I think when I approached them on that personal level, individually, after the meeting, it kind of made a connection with them."
Introducing value terms in need of alignment		"Public" "We are actively trying to educate the public on the illegal use of off road vehicles on public lands. And we're trying to come up with an approach that you know the public can say, they know what the rules and regulations are when they are either (a) on public land or (b) operating an ROV on public land. And one of the messages that we came up with was, you know, 'Keep your wheels on the street; use your feet.' And basically what that is, is, you know, vehicles are welcome on designated public roadways, but off of those roadways, we encourage people to hike or do other passive recreation, like, you know, hiking or mountain biking or other types of activities."

continued on next page

Table 6.1—continued

Ethos appeal	Explanation	Example quotations
Character		
Introducing value terms in need of alignment (continued)		"Manage" "One of the first things I did as the resource manager, my job is to ensure that the public lands that we've been entrusted with protecting are managed in a way that's conducive to the type of activity and specifically safe for public use." "Others were like, 'We never see you all down there; we never see the government down there. Who manages the area?' Well, that was to me identified as, well that's an issue; they need to know who, who's responsible for operating and maintaining the area. So I kind of noted that that was something that I needed to improve on."

© 2021 by the Association for Business Communication. Reprinted by permission of SAGE Publications

connect with Aristotle's concept of ethos as a credible, persuasive persona. When sincerity complements credibility, a speaker can persuade even more effectively than through appeals to credibility alone. To emphasize his eventual acknowledgment of sincerity's importance, Edwards mentioned several times during the interview the need to show how sincere he was in communicating about the issue of possibly closing Grey Cliffs; he understood the sensitive nature of the proposal to close the area given the previous tense interaction between the community and the Corps, and he wanted to create a relationship of sincerity and trust between himself as a Corps representative and the community. He states:

> I make sure that I conduct myself responsibly and professionally and ethically and that the person receiving the information can see that. How I'm treating them, how I'm providing information and relaying that information in a positive manner help take the edge off of the message that they're receiving. And maybe opens their mind, broadens their perspective a little, and then helps me form more of a connection when they see that I'm trustworthy and that I'm sincere in my desire to make a difference.

Part of being sincere also involves making sure the message is received by the intended audience in a way they can understand, including positive actions the audience can take part in. Based on the hostile community response at the first meeting, Edwards changed his focus and pivoted from discussing closure as one of several options to negotiating ways the community could help improve the area in order to keep it open, such as by participating in cleanup days, helping informally police the area, getting the word out to others via social media channels about what is allowed and not allowed, and helping to put up barricades and gravel to make clear what areas are for public use and to increase safe use of the area. By suggesting legitimate ways the community could be involved, Edwards attempted to portray himself as someone who sincerely wanted to work with the community. Edwards reflects in his narrative,

> But I hoped that I was pretty clear, I hope that I was more sincere and that I wanted to make a difference there for their families and for their kids that they could go there, and I thought that I did.

In Edwards's eyes, this sincerity would result in a long-term relationship that would have lasting effects on the area regarding sustainability and safety. These efforts would impact the area as well as relationships. Specifically, Edwards wanted to emphasize his role as a continued participant in maintaining the area through contributing resources; providing this material help was yet another way that Edwards hoped to indicate his sincerity through future interactions with the community:

> And I think that's what this all boils down to, in that, you know, I think they see, I hope they see, the sincere and willing participant that, um, I could bring a lot of resources to the table, and I'm willing to commit those resources to bettering the services that the public has at [Grey Cliffs].

While establishing sincerity was important to Edwards's creating a persuasive ethos with the community, expressing affinity with the audience was also important to convince them to change their behavior and co-construct an identity and relationship with them. The audience could have accepted Edwards's sincerity and, relatedly, his trustworthiness and honesty, but, based on the unpleasant historical relationship between the community and the Corps, Edwards needed to demonstrate that he was like the audience in some ways and that he and they valued similar things. Only then could Edwards begin to establish a long-term relationship with the community involved in this area.

Appeal to Affinity

A few ways Edwards demonstrated affinity with the audience were expressing common interests, reaching out to members of the community audience as a friend, and introducing value terms that indicated an ideal, shared value alignment. Edwards was basically trying to communicate to the audience, "I am like you," "We can be friends in these efforts," and "We have (or should have) common values, as a result."

Expressing Common Interests

One way Edwards tried to establish affinity with the audience was through expressing common interests. For example, he stated, "I like to hunt, too" and "I like to go off-roading, too," but he emphasized the need to ask permission and not do these activities in prohibited areas. His hope was that these common interests would help build a relational bridge between himself as a government representative and the community. He also tried to use his own experiences about asking permission as a model to encourage similar behavior from the community.

Reaching Out as a Friend

Another way Edwards made personal connections was by asking individuals to help him with his efforts as though these community members were friends. Friendship is one of the virtues Aristotle mentions in his *Rhetoric* that help create a persuasive ethos, and Edwards singled out meeting participants who were active in resisting his efforts in order to establish a friendship, hoping that the community members would be more likely to help him as a result. Through local conversations as he continued reaching out to the community, Edwards learned that, while some environmental damage was caused by people outside the community, other contributors were from within the community and objected vehemently to closing the area, in part because they used the area to practice for off-road vehicle races. Edwards states, "I just basically approached them and asked for their help." The regulatory power that Edwards sought, while seemingly his from the outset, required him to step back and enlist the support of the community to help him negotiate that power in order to accomplish social action that benefitted both the Corps and the community, as well as the Grey Cliffs area. Referring to two men at the meeting who were publicly very vocal against the closure and the restrictions against off-roading, Edwards states,

> I said, "Guys, we've got to stop this; this is something that we have to do; I need your help." And I think when I approached them on that personal level, individually, after the meeting, it kind of made a connection with them.

Others who used the area to prepare for off-road racing in other locations said, "Okay, I understand." These new friends then used their influence in the community to tell others about the new restrictions and therefore helped infiltrate the hostile community to motivate key members to change their behavior.

In addition to these relationship-building strategies, Edwards also recognized the need to build a bridge between Corps and community values that initially appeared to be very different. In order to do this, Edwards introduced value terms into his conversations with the community, not only to encourage values that both could agree upon but also to introduce the community to essential Corps values. Not being familiar with Corps values at first, the community needed to understand Edwards's values and perspective, a process that could then hopefully allow community members to support those values as well if enough commonality existed between them.

Introducing Value Terms in Need of Alignment

During his communication with the community, Edwards encountered a challenge in his relationship with them that he had to confront: the Corps and the community discussed what they valued using different terms and narratives. For example, even though the Corps and community ultimately wanted the same result—an accessible, protected, and revitalized lake-access area—their words for talking about these values were different. Reflecting their values, community members' discussions continued to be grounded in experiential and affective narratives of time spent in this beautiful area, such as stories of fishing, swimming, family gatherings, camping, hiking, and generally enjoying the remoteness of this natural area. As he heard these narratives, Edwards pivoted mid-talk to consider other options for Grey Cliffs besides closure, indicating his awareness of the public's desire to keep the area open. This was one value that the Corps and the community could share, but the audience needed to know more about Corps' values before any type of significant value alignment could begin. While preserving the area environmentally and promoting human safety were Edwards's ultimate goals, another goal he attempted to accomplish was aligning these conflicting environmental values between the Corps and the community. Edwards realized that only when the community accepted its role in protecting this area would members follow the Corps regulations consistently. It is in this space of rhetorical tension that Edwards continued negotiating his ethos with the community by appealing to affinity even more through potentially shared value terms.

Because words are value laden, they shape social and cultural realities (Burke, 1968). These words make up a particular lexis for specific interactions and contexts, a lexis that also represents values (Jaworska, 2018; van Dijk, 1995). Clarifying for the audience exactly what specific terms mean in technical discourse can generate goodwill between the speaker and the audience (Tillery, 2006), effectively developing the speaker's character even more. During the conversation surrounding

Grey Cliffs, Edwards used particular words that reflected the Corps' mission, culture, and values. The repetition of the words in Edwards's public presentation and in the interview transcript indicates the perceived (and in some cases real) relationship between an organizational communicator such as Edwards and the intended audience (Verboven, 2011; Walton, 2013), both necessarily involved in the dialogic process of negotiating value alignment and tangible, social action. This reciprocal relationship can be created and maintained in several ways, such as through shared terms (Allen et al., 2012)—in this case, value terms focusing on sustainability. The shared terms indicate ways the organization and community or stakeholders are working together toward a common goal. In Allen et al.'s (2012) work, shared terms between corporate training documents and employee interviews indicated alignment in sustainability values; the study authors were also able to examine areas where additional alignment *could* occur, based on terms that did not align. Organizational communicators such as Edwards will be viewed as more persuasive if goals and values are presented consistently to target audiences and can be evaluated as consistent by those audiences (Boyd & Waymer, 2011), contributing even more to a persuasive and sincere ethos. Only when this type of ethos exists will positive community and stakeholder attitudes grow (Pasztor, 2019; Shim & Kim, 2021, p. 393), aligning with similar public value orientations.

Among the many value terms that Edwards used, both in his interview and at the first town hall meeting, the two I discuss here to illustrate Edwards's attempt at value alignment are "public" and "manage." These two words reflect key functions of the Corps, which are to manage public lands for the use of future generations. While these terms seem relatively clear, their use represents how value terms can mean different things for different audiences, depending on their backgrounds and values.

"Public"

The word "public," for example, has a different meaning for the Corps than it would for non-Corps discourse community members. In general, public could mean "open to all," although the word "all" would need to be further defined and most likely would not be fully inclusive. It could also mean "for general use." However, for the Corps, the concept of "public lands," with "public" used as an adjective, relates closely to maintaining land for future generations, for the use of those generations; "public lands" entails lands "used by the public" but not "owned and maintained by the public," based on the public's relationship to the government and the Corps. These terms clearly reflect the Corps values of its required relationship and duty not only to its land but also to the public, from a government and management perspective. When Edwards used the term "public lands" with the community audience, he was using it in the context of its cultural history with the Corps, not really considering that the public may have been thinking of it as simply "lands to be used by the public." This disconnect might explain the community's previous use of the land; it was for their use, including all of the activities the public might desire to carry out there. This differing interpretation indicated an area in need of potential value alignment since the Corps and community understood these value words differently. Likewise, the community appeared not to understand the relationship of the public land and Corps management of it; they did not understand the value the Corps assigned to land management, as represented by the word "manage."

"Manage"

The general public did not even know who managed Grey Cliffs; some people had heard it was Corps property, but the Corps had been invisible up to this point. Individuals had had little or no interaction with the Corps before, a problem that could also contribute to the degradation of Grey Cliffs, specifically because of the apparent lack of ownership, management, and, therefore, enforcement. However, because the public did not maintain or own the land, they had essentially entrusted the Corps with protecting and managing these lands by default, even though it was an invisible entity. The term "manage" communicates the value that these lands mean a lot: they have been entrusted as a resource to a manager who will care for them for future generations. "Manage" relates to a basic value that the Corps communicates, since it conveys the need to care for the land, anticipating the current and future public's needs. Up until the point the community met Edwards, though, it wasn't really possible for the community to understand that Edwards or the Corps managed the land. The community had not seen any evidence of that process or relationship recently (because regulations and signage had been removed) and had not been thinking about that connection, even though the distant, cultural memory still remained of the Corps' creating the lake and dam.

Obviously, when Edwards used these words when speaking with the community, he did not do so to consciously instruct the community on the relationship among the Corps, the public, and managing the lands; he was simply using these words as part of the "Corps language" he had always used; this language included labels that embodied Corps values.

As Hartelius and Browning (2008) state, "In short, labels are a rhetorically constructed way of sorting organizational members into groups by assigning identity" (p. 27). This process is not always smooth or successful, though. The community did not realize the need for their values to align with those of the Corps by attempting to learn the meaning of these specific words from Edwards's value perspective. Yet presenting these value terms and discussing them with the community allowed Edwards, although subconsciously, to highlight for this community where some of the value differences lay—differences related to environmental sustainability values. Helping the community realize these values using an understanding of shared terminology is one way Edwards attempted to develop this negotiated, shared identity with the community, an identity that reflected the Corps and community as being similar and on the same team with the same environmental sustainability goals.

NEGOTIATING SHARED IDENTITIES THROUGH SHARED VALUES TO ACCOMPLISH SOCIAL ACTION

Once Edwards had introduced the community to Corps value terms in an effort to familiarize community members with them, a possibility opened up for Edwards and the community to begin negotiating shared identities through shared values; these identities and values were prerequisites for accomplishing joint social action. Sustainability efforts, such as the ones connected to the Grey Cliffs lake-access area, involve values of environmental protection, repair, and regrowth, as evidenced in Edwards's discussions about the Corps' mission statements, values, and goals. Although these values were legitimate ones, at first, the community appeared to value its ability to access this area more, a value that contributed even further to the polarization it was experiencing with Edwards. The community was resistant to change because its longtime use of this area was threatened. Edwards also needed to convince the community of the urgency of this message. If he could not motivate the community to act (in essence, to accept his value characterization of the area), it could revolt and resist the Corps' efforts, deepening the rift that already existed between the Corps and the community. This sensitive situation seemed ripe with risk (DeKay, 2011): while the values seemed clear, reasonable, legitimate, and worth the efforts of implementation from Edwards's perspective, the dynamics could yield even more conflict with the community than already existed (DeKay, 2011). Negotiating an invented and situated ethos (Mackiewicz, 2010, p. 408) would require Edwards to co-construct a shared identity with the audience to align differing value orientations and ultimately protect Grey Cliffs. Only then would the community see Edwards as a credible rhetor who could be trusted. Edwards had accomplished part of this process through his character development as he related to his audience members through efforts to display sincerity and affinity, but he needed to work with the audience more to establish common ground through shared identities and values. Only then could Edwards maintain a relationship with the community and realize lasting, positive change for Grey Cliffs.

Once he began to establish a relationship with the community through the use of relationship-building rhetoric and value terms such as "public" and "manage," Edwards encouraged community behavior that aligned with Corps values that included sustainability, trust, ethical behavior when carrying out Corps responsibilities, and relationship with the community. (See Appendix 6.A for a list of websites used to support/document the values Edwards emphasized.) All of these values related to the overall goal of compliance with government regulations. Edwards referred to these values as he spoke during the town hall meetings, and he also communicated them directly to the community more explicitly through texts. Edwards presented these values to the community through revised texts, different from the ones he had presented before when communicating in regulatory mode alone. Edwards hoped that these revised texts would reflect his desire to identify and negotiate with the community members and their concerns. These values were ones that Edwards wanted the community to accept as their own to accomplish a mutual value alignment with the Corps (see Chapter 7 for ways the community accepted and modified this list of values through the negotiation process).

Sustainability

As Goggin (2009) clarifies, "Sustainability is not a concept for preserving, conserving, or reserving the earth and nature solely for their own sakes, but also for their continuing benefit to human society" (1). Throughout his oral communication during his interview and with the public, Edwards referenced "future generations" that would benefit from changes that would conserve and preserve the Grey Cliffs area: "Well, as a resource manager, my primary mission is to ensure that the public lands, that the public has entrusted with us to protecting, are preserved, maintained for future generations. That's basically our mission." This mission is accompanied by certain objectives, which Edwards didn't specify, but he indicated that he is rated on how well he works toward meeting these objectives each year. Rather than being penalized by Grey Cliffs' notoriety, his performance is rated on how well he implements the Corps' objectives and mission. However, going beyond this Corps mission and the values it conveys, he also takes on this mission as a personal responsibility: "I didn't do this [suggest closure] to . . . gain any fame or notoriety; I did it because of my personal . . . responsibility as an individual and a basic human right; I mean it's to ensure that those lands are there for, you know, future generations, and I take that responsibility seriously." Complementing this value of sustainability are the Corps values of *trust* and *entrustment*.

Trust and Entrustment

The values of *trust* and relatedly *entrustment* revealed themselves through Edwards's interview and oral communication. As a Corps representative, Edwards believed that the Corps as a whole and he as an individual are entrusted with the public lands that it owns and maintains for future generations: "My job is to ensure that the public lands that we've been entrusted with protecting are managed in a way that's conducive to the type of activity and specifically safe for public use." While this value sounds abstract, at Grey Cliffs specifically, the current activity taking place did not coordinate with preserving or maintaining the land, and the resulting degradation certainly did not yield an area that was safe for public use. The values of trust and entrustment imply that the organization and its members are worthy of this trust and carrying out the Corps mission. The community was not going to believe in or accept these values otherwise and needed to see that Corps values were connected to ethical behavior as Edwards carried out tasks aligned with the Corps mission. For the community, Edwards was the "face" of the Corps—the only tangible representation of this organization that they could see. Edwards's ethical actions, therefore, were essential to establishing and illustrating Corps values and trust at an individual level.

Ethical Behavior and Completing Tasks

Directly related to the trustworthy persona that Edwards tried to create through supporting ethical Corps mission statements and values was the *action* that Edwards took to demonstrate trustworthiness and sincerity through what Edwards described as ethical behavior. As the ancient rhetorician Quintilian stated, an orator is "a good man [*sic*] speaking well" (Golden et al., 2011); Edwards may have been speaking well at the first meeting but needed to support those words with action to prove his character and intentions, especially with such a hostile audience. Edwards needed to demonstrate that his ethical behavior in completing tasks was a Corps value and his value as well. This process would necessarily take time.

Edwards stated his hope that the community could see that he was a person of his word, and, while Edwards did want the community to know he had the authority to close the area if he wanted, he also wanted everyone to know that other options were available and that he was willing to support the community through compromise. In thinking back on the opposition he experienced at the first meeting, he reflects, "Like the first meeting, I thought I was pretty open that I don't want to close the area." Only by following through in relationships with community members was Edwards finally able to convince the community that he was on their side as well with continued, collaborative, ethical, supported action.

Relationship with the Community

Edwards also valued a sustained relationship with this community. In addition to reaching out to specific community members and asking for their help in enforcing regulations at Grey Cliffs to reduce the crime and environmental damage occurring there, Edwards also enlisted their help in speaking with local businesses that might sponsor cleanup efforts and provide materials to help fortify the area in an attempt to encourage community involvement on a larger scale. Based on interviews with Edwards and one of the community members who went with him to speak to the XYZ Company, I learned more about how Edwards presented materials related to the needs at Grey Cliffs. At Edwards's request, Tom, a community member, accompanied Edwards to provide firsthand accounts of his experiences at Grey Cliffs and to verify indications of support he had heard from other members of the community. The proposed collaboration between the XYZ Company and the community was a success, and the company provided funds for gravel and barriers. This corporate–community interaction is just one example of the relationships Edwards valued and fostered as a Corps representative.

While Edwards was regulated himself by the discourse, mission, and objectives established by the Corps, he adapted these elements and framed them within the context of Corps values, which he then communicated to the community, whom he hoped to motivate. In this sense, Edwards learned the value structures of his governmental organization and then took ownership of them (Giddens, 1984; Simmons, 2007; Tillery, 2019, p. 31) in an attempt to accomplish his own communicative goals as a social actor. This process also entailed learning and adapting to the culture of this community, introducing Corps values in ways the community could accept and reciprocate back through a shared identity needed to accomplish common goals. This process took time, and these values presented themselves in the form of several texts that Edwards implemented using a variety of media to address different audiences in an attempt to align Corps values with community values in order to keep Grey Cliffs open. These texts contrasted with the texts Edwards first presented at the "regulatory rhetor" stage: they now reflected Edwards's sincere character and desire for affinity as he collaborated with the community audience, and these texts also reflected Corps values of sustainability, trust, ethical behavior, and continued community relationship. Far from being final products, though, these texts indicate their unstable, changing potential as Edwards attempted to address Corps needs as well as community needs, which would necessarily evolve and change over time. Ultimately, these texts addressed community concerns: Edwards hoped that the community would be persuaded to act, based on this growing evidence that he was willing to negotiate with it.

TEXTS THAT REFLECTED EFFORTS TO NEGOTIATE RELATIONSHIPS AND ALIGN CORPS AND COMMUNITY VALUES

Edwards communicated Corps values to the community during later town hall meetings through maps, rules and regulations, email messages, and stories. While these texts still presented the seriousness of Grey Cliffs' degradation and the need to preserve the area (Corps values had not changed), Edwards also confirmed his willingness to build a relationship with the community as they followed Corps regulations through these texts in an effort to co-construct an ethos and identity with the audience. He also clearly addressed community concerns in these texts, indicating an effort to address community values revealed in these concerns. The community responded in varied ways to these relationship efforts, indicating the dynamic nature of the effects of Edwards's discourse, as, in some ways, Edwards attempted to groom the community discursively (Pilger, 1998) to respond in ways that aligned with Corps goals. Importantly, these "texts are a juncture between regulation and agency, the technical and the social, and the organization and society" (Faber, 2007, p. 216); specifically, they indicate how these environmental efforts were continuing to be shaped by discourse (Tillery, 2019, p. 77). These texts clearly reflected Corps values as well as the potential for community agency. As such, they exemplify many complex dynamics of Edwards's work with the community as both worked together to resolve the Grey Cliffs conflict.

Map

Once Edwards began negotiating with the community through a relationship with them, he presented a different map at a later meeting (in contrast to the one presented at the first meeting that seemed to highlight consequences for the community) that laid out a developmental plan for governing the area with some restrictions. This map presented "Phase 1" details including placing cables and barriers, designating one trail entrance to allow land rejuvenation, and assigning primitive camping areas.

[GREY CLIFFS RECREATION AREA] DEVELOPMENTAL PLAN
PHASE 1, 2018–2019

Barriers and cables preventing all motorized traffic except in designated areas.

Designated trail entrance. All other non-authorized roadways will be allowed to return to natural state.

Designated primitive camping area at the southern section of the area.

No camping will be permitted in any other areas.

No vehicles permitted in "Tent Only" camping area.

Primitive RV camping permitted only in designated area. RVs may not block parking area or launching area.

At the bottom of the map, red text stated, "Primitive Camping Permit MUST be obtained prior to camping. Please call the . . . Lake Resource Manager's Office for more information at [phone number]." The map also included a legend and emphasized "NO MOTORIZED VEHICLES" in all capital letters across three of the forest/land areas on the map while also labeling what areas coincided with permitted activities, as indicated on the legend. The map also included an overview of the Corps regulations for the area:

NOTE: All rules and regulations will be enforced as according to Title 36,

Firearms or other weapons are prohibited.

Motorized vehicles off authorized roadways are prohibited.

Camping is permitted only in designated areas.

Fires shall be confined to camping designated areas.

All state & local laws/ordinances shall apply.

In contrast to the first map that highlighted the uniqueness of the area and its remoteness that would require further travel to access the lake, based on imminent closure, this new map officially represented the

Corps' negotiated action that would allow the area to be used while still remaining in compliance with the Corps' rules and regulations.

When this second map was distributed, the community knew that Edwards was willing to negotiate to keep the area open under certain restrictions such as installing barriers and designating specific camping areas. This map provided a framework that included the development plan, the rules to be followed, as well as the map and legend to indicate visually where certain activities would be allowed and where they would be prohibited. The map, while clearly emphasizing Corps regulations and restrictive boundaries, also incorporated public and community activities by clarifying their inclusion. Because "written communication can be harder to ignore than verbal" (Moore et al., 2021, p. 19), this map documented a significant agreement between the Corps and community members. Essentially, this map, with its additional information and changed design, invited "the viewer's active participation" (Eichberger, 2019, p. 17) and potential agency (Jones, 2016) that framed a positive way to help resolve this crisis from the perspective of both Edwards and the community members. Based on meeting observations and field notes, I concluded that the community was receptive to this revised map, since it reflected Edwards's desire to work with them. Community members seemed encouraged by being able to read a physical, tangible document that verified graphically, geographically, and textually that Edwards was willing to work with them in continuing to use the area so that community members could still participate in activities they valued. This example illustrates Edwards's dynamic, "environmental ethos" (Eichberger, 2019, p. 18), which "recognize[s] unintended exclusions and silences in our own research and practice" (Eichberger, 2019, p. 18) to document others' participation within environmentally sensitive contexts such as these community members' relationships with Grey Cliffs.

While this map appeared in hard copy form and seemed relatively simple technologically, scholars recommend that one way to encourage public engagement is through more technologically sophisticated "deep mapping" strategies, which "dig deeper into the complexity of places and use the map as a communication tool to explore cultural issues that shaped an area, changes to the ecology over time, geographic data, political associations, and the many narratives that develop a sense of place" (Butts & Jones, 2021, p. 8). These maps can also include "locative storytelling experiences that help users connect to environmental justice issues on site" (Butts & Jones, 2021, p. 8). By incorporating storytelling and more local, cultural information into maps such as these, organizational communicators and policy creators can help those interacting with environmentally sensitive areas to "visualize connections that seem remote from each other" (Eichberger, 2019, p. 18), such as the relationships among humans, animals, and nonhuman environmental elements. These connections "rall[y] a collective of humans and nonhumans to care for each other in an ongoing way," as Olman and DeVasto (2020) emphasize in their discussion of creating environmental risk visualizations that "[empower] . . . the most vulnerable agents" (p. 16).

Ideally, these more participatory technologies would create more of a dialogue between environmental communicators and their audiences, such as through the collaborative citizen science projects that Olman and DeVasto (2020) refer to in their work (p. 21). Incorporating community members' experiences into maps such as these contributes a more "affective" (Butts & Jones, 2021, p. 14) dynamic to maps, in contrast to static versions that don't seem to take into account the evolving relationships taking place among all. These dynamic representations of geographical spaces such as Grey Cliffs could create a more inclusive, multimedia visual of the activities taking place there in many different, more meaningful ways.

Rules, Regulations, and Sociogeographical Context

While some members of the community obviously disrespected signs displaying Corps rules and regulations in the past and could possibly continue doing so, the community members wanting to make a change received the list of rules on the revised map and expressed interest in abiding by them, wherever they appeared. To this community, such a difference between the first and second maps illustrated hope and evidence of Edwards's wanting to help the community, rather than his doing simply what was easiest and closing the area. As Tom, a key town hall meeting participant, reflected in his interview, "You know, then [Edwards] done a big 190 flip, you know, like from wanting to shut it down that one day to like, you know, he's calling me like, wow, you know, we want to do this, this, and this." To Tom, the revised plan and Edwards's reaching out to him to collaborate communicated a clear message that Edwards was willing to negotiate and had changed his mind from the original plan that appeared to exclude the community, even though essentially the Corps rules and regulations appeared to be the same. Edwards's connecting a willingness to collaborate with creating the revised map that included the rules and regulations (appearing also on the signs) appeared to make a difference in this community's willingness to accept and abide by them. The revised map paired the regulations with the activities accessible to the community so that the two appeared to be equal, rather than the regulations' dominating the community's wishes.

Email Messages

In preparation for another meeting that Edwards could not attend, he emailed a list of action items to another community organizer to discuss at the meeting in his place. This email message included descriptions of the first and second phases of recovery, tasks to be completed, and equipment/resources needed to accomplish these phases, such as the following:

- Repair road and keep from washing.
- Grade shoreline for improved access (need help from county road department).
- Delineate trails.
- Replant bald cypress and other native [name of state] species.
- Initiate reforest improvement plan.
- Pick a day for community clean up and work day. Corps will bring material (posts, Sakrete, signs, cable, etc.).

The timing of this document and the revised map is important: they were given to community members a little more than 5 weeks after the first meeting, during which the community had displayed such hostility and resistance. The relative timeliness of these documented plans and negotiations from the Corps indicated Edwards's willingness to help quickly at a very sensitive time in this crisis. These documented plans also served as "proof" that the Corps was indeed willing to help, even to the point of supplying time and labor to the conservation and cleanup efforts. Based on the plans revealed in this message, the community could see that Edwards's pledges were more than "just talk" or empty words; he was willing to contribute materials and funds to help improve the area. Edwards's timeliness in acting also conveyed a caring concern for community members; he was continuing to acknowledge the urgency of the situation and his willingness to follow through on his negotiated action with the community.

Stories

In a community that valued its own stories, it would take a while before the Corps narratives would be transformed into anything correspondingly positive, and, realistically, community members might not ever view the Corps in a fully positive light. As a beginning effort, though, Edwards and the community would need to develop new narratives and perhaps superseding counterstories that reflected a more positive relationship of working together, rather than remembering only the old stories. At the beginning of the conflict, it didn't seem possible for the community to find anything positive in past Corps narratives. Even though these narratives focused on mutually negative evidence relating to the conflict, these stories were valuable texts that currently reflected Corps regulations that had been transgressed (from the Corps' perspective) and community members' land ownership and history that had been lost (from the community's perspective). In the meantime, the community would need to rely on the documented plan for negotiation and rejuvenation of the area as a gesture of good faith to preserve the beloved area, a beginning of more positive actions that would contribute to new stories in the future. While the community and Edwards did need time to develop new narratives, those did eventually begin to emerge. For now, the community heard Edwards's narrative of projected progress for Grey Cliffs and attempted to accept it based on the other textual evidence it had received from him.

With new identities and ethos continually being negotiated during the evolution of events at Grey Cliffs, agency developed in renegotiated ways as well. The "faceless" Corps with its impersonal, regulating narratives evolved into the face of Edwards as a concerned, sincere community supporter who took cultural context into consideration during continued negotiations because, ultimately, he realized that only through a genuine relationship would the community be motivated to act. Edwards realized the need to communicate and work *with* the community to solve this problem rather than dictating solutions that remained disconnected from community culture, their world (Latour, 1993; Salvo, 2006, p. 234). This evolution process was far from following a continuum, though, as conflict could reemerge at any point during this negotiation process. With that being said, the community, somewhat sobered and shocked by the possible and perhaps inevitable, repeated, Corps "hostile takeover" of Grey Cliffs, valued their newfound power in helping manage and protect the area. While realistically the Corps still maintained an upper hand, legally the community seemed content with their acknowledged and respected role in the sustainability process. The community portrayed a different and more positive view of Edwards as well.

A GROWING, SYMBOLIC CAPITAL OF TRUST

Essentially, Edwards needed to continue conveying a trustworthy persona and develop a trustworthy reputation in order to negotiate a collaborative ethos with this community necessary to accomplish the needed joint social action. Edwards had to adapt his narrative to address resistant community members. While not organization employees, these community members were nevertheless essential participants in these environmental protection efforts because of the clear role of the public within Corps efforts: part of the Corps goals and mission related to preserving these lands for public use. Edwards needed to develop and maintain a "capital of trust" (Bourdieu, 2007, pp. 185–186) that he could exchange with the community while everyone waited for the progress of these efforts to be realized. This capital of trust amounted to cultural capital (Bourdieu, 2007, p. 187) that Edwards needed to develop and negotiate with these community members who wanted a positive resolution to this conflict but who hesitated to trust a member of the government organization that had built up a capital of distrust with this community in the past. Through his willingness to let go of some of his "authoritarian control" in his role as Corps resource manager, Edwards began opening up the process for procedural justice to occur, through "true participatory contributions by communities" (Walton et al., 2019, p. 39).

As Edwards pivoted to connect with this audience through sincerity and affinity, he and the community were able to begin establishing relationships to a degree: the dominating symbolic/cultural capital that Edwards conveyed at the beginning of this discourse turned to an exchange of social capital with the community eventually through an effort to align values as well as ethos. Through a capital exchange facilitated in part by narrative, these aligned values and ethos would lead to jointly accomplished social action on behalf of Grey Cliffs.

Developing trust in Edwards was a key part of this process of attempted value alignment and collaborative, social action; without aligned values, the community could not trust Edwards to do his part in helping the community keep Grey Cliffs open. Aligned values between the Corps and community were an essential prerequisite for the joint construction of agency between the Corps and community. One value the community and the Corps could ultimately agree on was that Grey Cliffs needed to be cared for in order for it to remain protected and kept open for public use; that was what both the Corps (based on its mission and values) and the community (based on its values and experiences) wanted. Agreeing on this one value was the beginning of creating more opportunities for social action and generating trust between the Corps and community. With this potential in mind, Chapter 7 presents the results of and community reaction to Edwards's ethos-negotiation efforts.

KEY RECOMMENDATIONS FOR TECHNICAL, PROFESSIONAL, AND ORGANIZATIONAL COMMUNICATION AUDIENCES

- Attempt to learn as much about community values as possible before initiating contact with local communities about sensitive issues.
- Introduce organizational values to targeted audiences while ideally also making connections to community values.
- When possible, incorporate as much location-based context into creating maps to emphasize cultural-historical context to increase audience inclusion and participation. As Butts and Jones (2021) indicate, "Digital mapping methodologies need new, location-based design approaches that can better account for the many layers of meaning at work within any place" (p. 5).
- Consider "deep mapping" (Butts & Jones, 2021, p. 8) as a way to emphasize the rhetorical context of maps and what they represent for geographical spaces and those interacting with them, including the power structures in effect.
- When communicating, consider who might be excluded and what information might be left out that could aid diverse community members and even potential, future audiences, if that information were to be included.
- Consider also any documents that "might unwittingly expose members of the community to undue scrutiny and function as an oppressive activity" (Walton et al., 2019, p. 116).
- Clarify community members' roles in environmental improvement efforts through specific texts, such as maps, that uniquely address community activities.

KEY RECOMMENDATIONS FOR ENVIRONMENTAL SCIENCE AND PUBLIC POLICY COMMUNICATION AUDIENCES

- When communicating with the public, be willing to pivot from the original plan if the audience responds hostilely or unreceptively to the message.
- Consider adding narratives to the process of conveying scientific information (including document creation) to clarify environmental impacts on local communities (Stephens & Richards, 2020, p. 7).
- Consider the relationship between scientific reports and the creation of public policy; this relationship can "implicate larger political and cultural systems" (Graham & Lindeman, 2005).
- Ideally, identify community values that connect to the environmental issues under discussion, and connect with those when creating public policy affecting those local communities.
- Assure local communities of their worth in connection to environmental issues, listen to their concerns and stories about impacted areas,

and attempt to incorporate public feedback into final policy decisions, recognizing "built-in and historical power relations that oppress individuals and social groups" (Walton et al., 2019, p. 116).
- Identify ways to increase community engagement to facilitate public policy implementation with community support, such as through citizen-science projects (Olman & DeVasto, 2020), technological tools that could increase the impact of cultural-historical information (Butts & Jones, 2021), and storytelling (Eichberger, 2019).

APPENDIX 6.A

Sites consulted for identifying value words for the U.S. Army Corps of Engineers and Sustainability:

Electronic Code of Federal Regulations. https://www.ecfr.gov/cgi-bin/text-idx?c=ecfr&SID=06f812f26e7ed9f364bb87944757b912&rgn=div5&view=text&node=5:3.0.10.10.9&idno=5#se5.3.2635_1101

Covering Part 2635 of the Code of Federal Regulations specifically, this electronically available document provides details on all of this regulation's sub parts, including general ethics advice, honesty, financial conflicts of interest, and use of official time. Part 2635 begins with "Basic obligation of public service." Many sub parts of this regulation address values and ethical behavior supported by the Corps as a governmental organization.

Electronic Code of Federal Regulations. https://www.govinfo.gov/content/pkg/CFR-2012-title33-vol3/pdf/CFR-2012-title33-vol3-part326.pdf

This section of the Code of Federal Regulations applies specifically to Corps of Engineers enforcement of regulations. It covers topics such as handling unauthorized activities on Corps properties, including various strategies for addressing them. This part of the regulations also addresses noncompliance issues and legal action that may need to be taken, including the process of administering penalties. This document illustrates the authority that Corps of Engineers managers, for example, have in enforcing Corps regulations, as established by the Code of Federal Regulations.

United States Army Corps of Engineers. https://www.usace.army.mil/Portals/2/docs/EP1165-2-316.pdf

This document provides specific guidance about rules and regulations applying to items of concern for the Corps on its properties, including regulations about the use of vehicles and activities such as swimming, picnicking, camping, and hunting. Regulation 327.2.C refers specifically to operating unauthorized vehicles on Corps property, and 327.2 focuses on vehicle use more generally.

United States Army Corps of Engineers. http://www.usace.army.mil

The main site for the U.S. Army Corps of Engineers, this page provides information on and links to things like Corps missions, locations, careers, and media. It also includes efforts related to current events, such as weather-related impacts on lakes, rivers, and dams, as well as other issues such as Coronavirus response.

United States Army Corps of Engineers. https://www.usace.army.mil /Portals/2/docs/Contracting/UAI_FINAL_UPDATE1_HCASigned _07APR14.pdf

A pdf accessible through the U.S. Army Corps of Engineers main site, this document covers topics such as the federal acquisition regulation system, including policies about conflicts of interest, publicizing contracts, and acquisition planning. This 330-page document also covers ethical concerns such as acquiring properties owned by nonprofit organizations employing people who are blind or severely disabled, privacy and freedom of information, applying government acquisitions to labor law, and handling protests, disputes, and appeals.

United States Office of Government Ethics. https://www2.oge.gov /Web/oge.nsf/Resources/5+C.F.R.+Part+2635:++Standards+of+ethical +conduct+for+employees+of+the+executive+branch.

This site provides resources for employees of the Executive Branch seeking information on codes of conduct, such as "The Fourteen General Principles," "Summary of the Standards of Conduct," and "Compilation of Federal Ethics Laws."

7
NARRATIVES OF JOINTLY ACCOMPLISHED SOCIAL ACTION THROUGH ALIGNED VALUES
The Negotiated Resolution

> *"Uh, it's a beautiful place. And it finally now, and now we got an opportunity to make it even nicer, and so it's time to jump on it."*
> —Paul, community member and owner of the general store where the town hall meetings took place

> *"Now it's going to stay open, and we're going to make it, you know, the community is going to make it nice, you know."*
> —Lee, community member whose farm borders Corps property adjacent to Grey Cliffs

> *"Since then, the whole dynamic of [Grey Cliffs] has shifted. The paradigm has shifted. The support of the community has made to me the difference."*
> —David Edwards, Corps of Engineers resource manager

Thus far, this case study reveals the need for some type of negotiated ethos, narrative, and discourse frame based on common values all could agree upon. Ultimately, what was needed in the Grey Cliffs community was innovation: the current conditions and activities no longer worked for Grey Cliffs, and Edwards, in response, found ways to strategically communicate his organizational message for needed change. As researchers have noted, messages for change must be accepted as legitimate by the participants involved, and this process often includes the use of persuasive, rhetorical strategies (Alvesson, 2004; Anand et al., 2007; Gherardi

& Nicolini, 2002; Smith et al., 2020; Suddaby & Greenwood, 2005). This chapter presents the results of Edwards's efforts to negotiate his ethos, including credibility and character, with this unique, rural audience. I begin this discussion of results with changed narratives and stories from the community; these appeared to prompt changed community actions but also co-occurred with them. I also incorporate and analyze texts the community created that document changed community behavior, supporting the changing narratives and stories. Edwards affirmed these changes through his own narrative. I conclude the chapter by reflecting on the aligned, co-constructed ethos and value frames between Edwards and the community.

DEVELOPING REFRAMED NARRATIVES AND STORIES

While not unified or essentialized in their themes, past narratives and stories had not been positive about the community's relationship with the Corps. That past narrative necessarily impacted the narrative that began developing when Edwards suggested that Grey Cliffs be closed; it also was grounded in negativity, even before it began. The stories and rhetorical personae of past collaborations and communications necessarily impact current perceptions and understandings of events and possibilities. As Faber (2002) writes, "The stories we tell interpret and create meaning out of the changes we experience" (p. 21). The stories told in this case study indicate how all participants were attempting to create meaning and find ways to address the changes (Weick, 1995) taking place in an area that had already witnessed so much change to begin with, as the Corps constructed the lake in the mid-20th century. To understand further the power of stories here, we have to acknowledge the powerful role that stories have in negotiating change. "Stories broker change because they mediate between social structures and individual agency" (Faber, 2002, p. 25). The strong, regulatory, social structures that the Corps had put into place were a strong past influence in this community. Throughout the present changes taking place, Edwards and the community told stories that verified the changing interactions between the Corps and the community: these stories served to counter the past stories of the Corps, when resisting the Corps' social structures did not work out well for the community at all, as many landowners lost their homes and farms. These "kairotic antenarratives" include "intuitive understanding of local conditions in order to be able to capitalize on fleeting opportunities and brief openings in dominant narrative strategies" (Herder, 2015, p. 359). This intuitive understanding led the community to tell stories about how they would like to see the Corps, the community, and Grey Cliffs in the future. Having internalized the structures of previous Corps actions and communication, the community now attempted to change these structures through new narratives and stories, similar to what Giddens (1984) discusses when presenting agency as a process of internalizing structures and then acting as agents within them.

Essentially, what began as a narrative framing of compliance and regulation of the community's actions, based on Edwards's personal experience visiting Grey Cliffs, observing the litter, erosion, and drug use there, as well as documented criminal statistics from the County Sheriff's Office, turned into a "counterframing effort" (Waller & Conaway, 2011, p. 85), co-constructed by the community, that emphasized instead how the Corps was willing to help the community through donations of time and materials, for example, to improve the area and keep it open. When the community recognized Edwards's alternate framing strategy that avoided the implication of involuntarily accusing the community of crimes, members began exhibiting a willingness to help and collaborate with Edwards. Based on community members' continued stories about the changing character of Edwards and his characterization of the community (in essence, not criminals but people who had gone just a little too far in enjoying using the land), people wanted to start helping and working toward a common goal to decrease crime and environmental damage.

Once Edwards demonstrated acceptance that the community members at the meeting were indeed not the ones responsible for the crime, the framing of the issues and conversation could then focus on what the community could do to help reduce crime, such as informal surveillance of the area, enforced by the placement of cameras at Grey Cliffs, and community observations that they committed to communicate to Edwards via cell phone. Notably, this reframing process was developed and negotiated discursively "from all sides" instead of being oppositional in nature: Edwards reframed the developing narrative about this conflict and indicated a willingness to compromise, dialogue, and collaborate, and community members did the same. Far from being the judgmental and blaming narratives that both parties initially spoke in opposition against each other, the new, reframed narratives focused on change, possibilities, and collaboration. While the community probably didn't realize the power they actually had to negotiate with the Corps, based on the Corps' responsibility to interact with and notify the public of changes occurring with the lands it used, that agency and power were realized through this negotiation process with Edwards: the community drew upon the bargaining power they had to contribute to future narratives about the Corps. The reframed narrative became "symbolic capital" (Bourdieu, 1986, 1987, 2007; Faber, 2002, p. 156) that the community used to negotiate that agency and power. This developing, common narrative was a prerequisite for change to take place in this situation. This reframed narrative contained stories the individual community members narrated that reveal many different themes, such as changing collaborative actions, changing perceptions about Edwards, and working toward a more positive future. These stories were also key to addressing social and environmental justice concerns existing within this community so that all could participate in enjoying this land while ensuring environmental preservation. These stories indicated community agency through changed narratives as well as actions.

Beginning Stories about Changing, Collaborative Actions

Before the community could create a narrative supporting collaboration with the Corps, active community members, such as Tom, needed to acknowledge that a problem even existed. Tom, as a representative of the community, eventually acknowledged that environmental change needed to happen at Grey Cliffs; before, the community as a whole seemed resistant to acknowledging that anything needed to change. Tom states:

> You know, and I understand part of the environment, like he [Edwards] said needs to be regrowed and, you know, jeeps and stuff, and people had beat it down. See, that's the thing. Everybody in jeeps it wasn't from here that was going down there; they was from like [name of] County, [name of] County, you know, [name of nearby city]. People was coming from everywhere, you know, "hey let's go down to [Grey Cliffs]," you know. 'Cause that's like that night he was talking about [Slippery Ridge], he said, he knew where that was at, he said, "but there's nothing can go, come up that." "Well, you're crazy man," I said, "I've been up and down that in a jeep many times." I said, "I've been up and down that in a jeep." "There's no way," [Edwards said]. I said, "yeah."

Tom acknowledged the environmental problems, especially ones caused by others from outside the community; however, he also subtly acknowledges his part in the problem at the same time, when he talks about his part in off-roading at a location near Grey Cliffs. Instead of simply admitting this, though, Tom's story adds evidence to a previous divide and disagreement he had with Edwards about the types of activities occurring in the areas surrounding Grey Cliffs, indicating how he ultimately agreed with Edwards about the problems. During this process, Tom assigns more responsibility to outsiders who were visiting the area. This example indicates the pushing and pulling dynamic of agency between Edwards and Tom as a community representative: he acknowledges a limited role in the problem while assigning more responsibility to hard-to-identify outsiders.

Once community members acknowledged the problematic actions they could change and accepted a relationship with Edwards, based on his character development, stories surrounding community and Corps collaboration could begin developing. These stories focused on cleaning up the area and visiting a local business to ask for financial support to preserve Grey Cliffs. As Tom considered the upcoming summer, he predicted that people might start trashing the area again, and he proposed to Edwards the idea of putting trash cans down at the lake-access area:

> You're gonna have a lot of people down there that's camping and, you know, hiking, the boats and stuff, so. There's probably going to be a lot more cleanup 'cause everybody don't pick their trash up, you know. People are, bad to say, but people are nasty. I mean, a lot of people are, they just, beer bottles, cans, whatever, the garbage is needing to be cleaned up. And that's what I told [Edwards], I said, you know, we need to set up poles with metal garbage cans with bags in it so they got somewhere to put their trash. I said because there's nowhere other than, they don't take it with them; it's on the ground. And he said that was a good idea, you know.

As the spring months arrived, the community advertised and hosted cleanup days to keep the area trash free and to maintain the improvements that had been made; these efforts connected with previous community promises about what it could do to help improve the area and keep it open. Stories about collaborating for these cleanup days and suggesting very specific strategies, such as installing the trash cans, connected to previous narratives in a way that proved the community had followed through with what it had said it would do, if the Corps kept the area open. Edwards was supportive of these ideas, and this support enforced the growing theme that the community could work together in small and more significant ways to accomplish collaborative goals.

Another story that Tom told that supported growing collaboration between the community and the Corps was the narrative of Edwards going with Tom and Dan to visit the XYZ Company and request help. Interestingly, as discussed in Chapter 5, Tom resisted preparing for this meeting and did not create a formal elevator pitch for it. To him, that seemed to be part of the resistance process against Norma, who insisted that Tom be prepared with one. However, as he and Dan waited in the parking lot of the XYZ Company, the two of them did do some discussion and planning that included Tom's encouraging Dan based on his previous professional experience. The two were formulating a plan for how they would all work together, and Tom admonished Dan to "get your notes together":

> And me and Dan, you know we had to sit there and wait on—me and Dan set there, you know, in the truck waiting on [Edwards] and his crew to get there, and uh, I told Dan, I said, hey, I said uh, you know, if you want to take notes, feel free to, and I said, you know, get your notes together and everything, I said, 'cause he [Edwards] was impressed with you the other day. He's like, well, he said, I worked for a ah I think it was in Alaska for a some oil company, and that's what he done, he come out, like oil conservation stuff, you know, and he had to take a lot of notes, and organize, went to a lot of meetings you know and covered a lot of them. You ought to know what we're talking about, you know, too, so that elevator pitch. So, you know, he was real good, and I said [to Edwards], "Well you go ahead and do whatever you need to do because Dan is real good with that." So, you know, yeah, and he's got [Edwards]. But Dan's like, "You know more about [Grey Cliffs] than I do; you've lived there." I said, "I understand, so we'll make a good pair." I said, you know.

Tom's story about how he and Dan worked together really was about establishing ethos and co-constructing an ethos with Edwards to accomplish these social, environmental goals. Edwards's ethos for this meeting originally was somewhat defined as the Corps resource manager, although there certainly would be an ethos-negotiation process during the meeting as all of these participants from different backgrounds communicated about these issues. It really was a dynamic setting with the XYZ Company's management, such as the CEO and plant manager, plus Edwards, Dan with his previous lobbying experience, and Tom narrating his personal stories. While this story was all about these participants working together, Tom clearly allied himself with Dan as a fellow community member by saying, "We'll make a good pair." Such statements indicate that while Edwards and the community were beginning to negotiate a process of accomplishing social action together, the differences among these participants, based on values, experience, expertise, and positionality, all would influence the communication process in the dynamic, ethos-development process. This type of negotiation illustrates the collaboration that can occur when government experts, for example, aren't the only ones viewed as possessing expertise (Tillery, 2019, p. 107). Rather than community members' needing to acquire similar expertise in order to participate in discussions, in this case, localized community experience with the contested space served as valuable capital for ethos negotiation.

Contrasting with Aristotle's concept of an individual rhetor, Reynolds (1993) highlights this potential for the co-construction of ethos:

> Ethos is not measurable traits displayed by an individual; rather, it is a complex set of characteristics constructed by a group, sanctioned by that group, and more readily recognizable to others who belong or who share similar values or experiences. The classical notion of ethos, therefore, as well as its contemporary usage, refers to the social context surrounding the solitary rhetor. (p. 327)

Tom reflected on the social context of this important meeting with the XYZ Company; he narrated how the meeting went, emphasizing the process of how all of these different participants worked together to create a convincing argument for the XYZ Company to contribute to the conservation efforts. This example especially indicates the social context through which each meeting participant/rhetor helped construct each group member's communication. This meeting with the XYZ Company contained some texts from the first town hall meeting, during which Edwards had discussed the map of the lake area that he was using now. The dynamics of that discussion, though, had totally changed as this group worked together to persuade the XYZ Company's executives:

> And [Edwards] went through the whole thing showing them a map and how it looked and the Corps and how the water level and all, and he throwed his elevator pitch and it's like. And then they uh, he, they asked me and Dan questions, and Dan, he answered what he could, he said, I've not lived here a long time, he said, but now [Tom], and I told him, you know, about learning how to swim down there and that I was actually from [another local community], but we always come over here, you know, when we didn't go to [Rocky] Bottom, then we would come over here, so, we was over here a lot, too. And I said, told him what a great place it was, you know, and my kids go down there and swim and [y]eah, and I said, they have a lot of baptisms.

Each of the characters in this story had his own role illustrating the social collaboration taking place: Edwards had a formal elevator pitch that included maps and statistics about the water level. Dan had his experience with oil conservation to discuss, and then he passed the microphone to Tom, who could emphasize his personal experiences growing up in the area that stressed the values of recreation, spending family time together, and baptisms. Together, with all of these different rhetorical elements of credibility, experiences, values, pathos communicated through stories of the area, and the logos of statistics, the persuasive discussion apparently went quite well:

> But that was, that was all that was really said, you know, that they [the XYZ Company] wanted to help clean up, whatever needed to be done, and

they would invest money for gravel or rocks, or, you know, poles, or whatever, but that, that was, you know, and that's what [Edwards] went there for, to get their help.

Together, these efforts succeeded in convincing the XYZ Company to contribute materially and financially to these efforts. The seemingly diverse narratives and experiences worked to negotiate a complex, dynamic, far-reaching ethos that resonated with the XYZ Company executives.

Paul also confirmed this new, collaborative narrative that the Corps and community were now working together: "Yeah, yeah. [Edwards] even brought in machinery, I mean, it's working great. Uh, we've got a good, good communication, and a good plan going forward, I think." Importantly, Edwards wasn't looking just for company funds to sustain Grey Cliffs; the Corps contributed funds to the efforts, as well. The stories being told reinforced the fact that all were contributing to this collaborative effort.

Through these types of stories, the community began revealing its own character to Edwards, setting the stage for potential, continued ethos negotiation. Reciprocal trust was needed so that Edwards could see the community was dedicated to improving the area (in essence, that its values were aligned to some degree with Corps values) and could reasonably envision working with community members. For these community members, agency was constructed/negotiated through these narratives of ownership and compliance, through collaborative effort with the Corps. Beginning the agency negotiation process was difficult for this community, since "the problem of agency is the problem of acting within systems of decision-making marked by organizational, epistemological, and discursive complexity" (Grabill, 2006, p. 159). As the community continued initiating this process and working within these powerful, discursive structures, their stories also changed by highlighting changing perceptions about Edwards.

Stories About Changing Perceptions of Edwards

Another emerging story theme reflected a change in community members' perceptions of Edwards and his involvement in the Grey Cliffs conflict. At the first town hall meeting, the community couldn't believe that Edwards wanted to close the area. Edwards had stated in his interview that closing the area was just one option and that he was, indeed, open to other options to maintain the area, but the community couldn't seem to get beyond the seemingly accusatory stories of crime, including drugs and violence, that Edwards narrated. To the community, these were "bare-faced lies." How could the community argue against these lies, though, in the face of this "evidence"? However, once Edwards indicated willingness to work with the community, community members seemed much more welcoming of Edwards. For example, Tom had voiced his surprise that Edwards was wanting to brainstorm ideas with him about maintaining the area, and Felicia said, "There was so many people here, and he seen the genuine concern of the people that didn't want it closed down." Tom especially was taken off guard by this apparent sudden change in Edwards, but he was very pleased and wanted to begin the agency negotiation process by talking with Edwards about these differences in opinion about what had been going on at Grey Cliffs:

> A lot of that, a lot of that, me and [Edwards] talked about that. And I told [Edwards], and I said, you know, I know, you know, just like when we was in the parking lot at [the XYZ Company], I said, I understand that no matter whatever public park you go to, whether it's [another lake-access point] or [unintelligible], I said, there's probably drug dealings going on right under your nose, and you don't know it. So if you want to call that crime, okay, but, you know, and that may have went on down there [at Grey Cliffs], but as far as any killings, I'm telling you, it was rougher back then when all of us was younger.

These conciliatory conversations that developed between Tom and Edwards went far beyond the lies Tom insisted Edwards was promoting at the community meetings. Tom was even beginning to recognize and acknowledge that perhaps drug dealings had been going on at Grey Cliffs after all.

Tom believed Edwards was ultimately "a changed man." He emphasized,

> Yeah. [Edwards is] a nice guy, I mean, I really, I got a better opinion of him now than I did have that first day, and he told me, he said the word, he said, "[Tom], that [the first town hall meeting presentation] was all scripted." Yeah. I've just, I said, "and I know who scripted it, too." And he goes, "Yeah, but I'm not mentioning no names," he said. He's different. To me, he's different from what he was that first night.

At the beginning of the conflict, the community had interpreted what Edwards said as a problem Edwards invented or at least escalated himself, as he described the seriousness of the issues facing Grey Cliffs and the need to close the area. However, he told Tom that someone, presumably Norma, had told him what to say at the first meeting, to guide him into a resolution that Norma and a limited number of community members wanted. As the trust grew between Edwards and Tom specifically, Edwards admitted to him that some of the presentation at the first town hall meeting was actually not his own. This admission provided Edwards with an explanation for his first proposal that had seemed so radical. Norma, then, became kind of a "scapegoat," a person for absorbing more of the community's distrust so that Edwards could gain more agency and, using his authority, begin to work with the community to accomplish change.

Paul confirmed this changed version of Edwards that Tom had narrated:

> Yeah, the Corps, in the first meeting, he wanted to shut it down. And then, uh, we had like [two hundred] community members here, I think. And I think that blew his mind. Uh, I think the, I think what was brought out that they didn't realize how major that place is to the community here. And how big this community actually is. I think that was a shock. But ah, once he found out what the community was willing and offering to do, then it made [him] change his ball game.

And Lee communicated this same type of change that suddenly took place that started to enable collaborative, negotiated action:

> Yeah, I think uh I think if you remember the first couple of meetings, um, the authorities were not very responsive; it wasn't until we had smaller meetings, and we got the people out of there that was arguing, that they actually started listening. And those who were speaking with some intelligence, like you know these are the reasons we want to keep it open, then, you know, the authorities began to listen and started agreeing that maybe we can make something happen.

Based on these community stories, the authorities were compelled to listen to these voices or risk negative publicity and resistance in an area that had already experienced much negativity. The community then acknowledged this change in the behavior of "the authorities" and accepted it as part of the agency and ethos-negotiation process.

Another changed perception of Edwards involved narratives about options for ways to keep Grey Cliffs open. These options indicated that closure was not the only option, according to Edwards, and that the community would have "hope" that positive change could occur to the area while still allowing community access. As Tom reflected,

> Because I talked to [Edwards] about [the community leasing the area], he said, look, he said, you take it over and lease it, he said, you got to put lights up, you got to have running water, you got to, you know, be able to make it like . . .

At this point, Edwards had become a trusted advocate of the community, someone who could provide guidance about the options available, such as the possibility of the community's leasing the area. In community members' eyes, Edwards had changed so significantly that some people agreed that ultimately the Corps should retain control of the area after all. As Tom reflected,

> So, you know, it was better off for the Corps to be able to still keep control of it, that the community still help, which I guess the community's more or less doing . . . all the leg work and the labor, and then, you know, the Corps [is] not; he's still bringing people down, but we're doing most of the free labor, you know, and that's, it works out good, so, I mean, you know, he's keeping it open, and he's helped it, you know, shut the 4-wheelers and jeeps down, and all that stuff. So far.

To Tom, accepting Corps control of the area, which at first was a key part of this conflict, was worth still being able to access this beloved, geographic space. He knew that this rural community could not afford to support maintaining the area if the Corps were to lease it to the county, for example. Tom's resignation reflects the overall sentiment of the community that the most important goal was keeping Grey Cliffs open and available for the public to use, in a way that would support environmental rejuvenation.

In addition to stories about changing perceptions of Edwards and changing, collaborative actions, community members also told stories about working toward a more positive future. Significantly, the community stories did not remain in the past, focused on the Corps land takeovers, and they didn't stop in the present with current successes. Instead, the community and the Corps wanted to continue these efforts by looking to the future, and they framed this discussion through stories that discussed future plans and efforts.

Stories About Working Toward a More Positive Future

A final story theme that developed, which in turn fed into a developing new narrative about Grey Cliffs, focused on working toward a more positive future. These stories about improving Grey Cliffs, continuing to work together, and continuing to hold meetings essentially functioned as types of self-regulatory tools (Heath, 1983, p. 185), tools that attempted to shape the future of the community and its relationship to Grey Cliffs.

Stories About Improving Grey Cliffs

Aside from stories of cleaning up the area, community members had some additional ideas about improving Grey Cliffs, making it more attractive for people to visit, and connecting to other natural areas. Community members had already stressed the fact that Grey Cliffs was important to keep open because from it people could reach other lake-access points and other areas bordering the lake; in other words, it was a convenient location for people to launch boats and kayaks for those wanting to fish and explore the area. But Tom, for example, had some more elaborate ideas about enhancing the area. These ideas were sparked during the meeting with the XYZ Company, which included another member of the community who was a "nature person," according to Tom:

> And one of them [an attendee at the meeting] is, a couple of the guys was, I call like nature people, you know, like to get out, and he rode, he was from east [name of state], and he rides, uh, bicycles, like trail bikes . . . , like mountain bikes and stuff like that. He said, you know, is there, 'cause [Edwards] said you think there's anything like a walking trail or a bike trail we could make, you know, that we'd put money in to invest in that. [Edwards was] like well, I said, well, you can actually go to [Grey Cliffs], and I said, I can walk around and take you all the way over to [Clearwater] Falls. I said, it's a good hike, I said, but I've been through there on a 4-wheeler, and um, and [Edwards is] like, really? I said, [Frank Byrd] has, too. I said, so I mean, you know, I said there's a way you could cut a trail; I said, you know, it would be on Corps property; it wouldn't be on nobody's property. And I said, you could cut a trail, and woodchip it and whatever you'd have to do to, all the way into [Clearwater] Falls. And I said, you know, it'd be a good trail for bikes or people to walk on, whatever, you know. They talked about that, and [the XYZ Company] talked about, you know, putting in money for that.

Here, Tom, Edwards, the XYZ Company, and the "nature person" were collaborating about ways to expand the area, to connect it to another state natural area that the Corps did not own, and community members showed interest in connecting Grey Cliffs to Clearwater Falls more closely, expanding access to both areas. Making this connection would physically connect two locations: Clearwater Falls already had established infrastructure as a state natural area, and Grey Cliffs was developing that infrastructure through collaborations between the Corps, community members, and the XYZ Company.

In his interview, Edwards clarified his vision for the area that coincided with what Tom had narrated:

> And then the long term is, you know, you see it changed environmentally and with restored trail going back to where [Lee's] place is and tying in to the rest of the government line and some of the, you know, low gaps and maybe making a trail that goes to [Top View] Falls and tying in to some of the other recreation areas and then marketing the, uh, through the County Chamber of Commerce and other Chamber of Commerces to ensure that the public and, you know, other entities are aware of the

recreation benefits that this area offers, I mean, no other place on the lake can you go to that you can access [three state natural areas]; I mean, canoeing and kayaking is probably one of the most popular, um, paddle industry in the nation, I mean, it's the hottest, and you look at these groups that are already out there looking for areas, and there's, um, what are they called, they're like recommendations, you know, there's um, there's a YouTube on camping at Grey Cliffs, there's a guy's, you know, give a thorough recommendation of the facilities that are available there.

Edwards was interested not only in changing the narrative about Grey Cliffs with the community—changing it from a narrative of crime and degradation to one of "recreation benefits"; he was also interested in changing the public narratives and stories about Grey Cliffs that extended beyond the community. This particular influencer Edwards was referring to warned potential visitors to Grey Cliffs about the trash in the area and the crime; according to Edwards, who had viewed the video, he stated in his YouTube video, "Don't leave valuables in your car, don't leave, you know, anything left in your car, basically, you know, it's going to get broken into."

Even though Edwards was interested in and concerned about Grey Cliffs' reputation and impact beyond the local community, which was probably Grey Cliffs' most visible, public reputation, he necessarily understood that the local community, who would be using the area the most, was also a vital group to consider when envisioning the future benefits of Grey Cliffs' changing dynamics.

Edwards continued his positive Grey Cliffs narrative by describing a place where not only the public (local and beyond the community) but also the environment would benefit from the changed perspective and efforts:

> I would say that it [the Grey Cliffs area] . . . has a bright future, I see an area that kids are going to be able to go to and learn from and experience, you know, the great outdoors.
>
> . . . We can lay out, um, a facility that will be . . . sustainable for the environment, and you can, you know, reduce, less runoff and less erosion issues which affects water quality there.

Edwards's stories about Grey Cliffs' future spanned many different populations and entities: the Grey Cliffs' physical environment, outsiders who were drawn to the area, the local community, the Corps, and local businesses willing to support the cleanup efforts.

These stories also extended into physical infrastructures that funding from the XYZ Company could facilitate, such as installing lighting equipment and security facilities, as Tom had recounted. While negotiating the process of what the possibilities were for improving the area physically,

Edwards was also very clear that the Corps was going to maintain control of the area. Despite the changes taking place that would involve a variety of different people—private citizens as well as a corporation—Edwards established his ultimate ethos authoritatively, based on his role as resource manager. However, Edwards was willing to negotiate his authority with community voices in an effort to work together, even though, technically, the community wasn't "officially" entitled to an opportunity to negotiate this authority. While the community may have been able to improve the area on their own, Edwards demonstrated the need for more people—more stakeholders—to be involved in this process. A greater investment by all would in effect distribute an "ownership" and investment in Grey Cliffs that could not be established as effectively through Edwards's efforts alone. Narratives about Grey Cliffs' future improvement, therefore, were enabled as well as constrained (Giddens, 1984).

Stories About Continuing to Work Together and Communicate
Another story that community members specifically told about the future focused on continuing to communicate with Edwards. Tom, the creator of the Grey Cliffs Facebook page, saw Edwards's "liking" the page as an indication that the page was working as an effective means for distributing information about Grey Cliffs:

> But I have a lot of, I've not had no negative comments other than people, you know, just told that they couldn't camp, or they couldn't come down there, that a lot of them wanted to know that they had heard that the community had it shut down, and I said no; that was hearsay. You know; I don't want them [to think they can't visit].... Well, [Edwards] even liked the page and made comments on it, so I mean, you know, he liked it, too.

As a result of the initial conflict, many people from outside the community assumed that the area was closed and didn't want to attempt to visit the area. Part of the Facebook page's role was to provide correct information about exactly what was going on. Edwards's sanctioning of the page reassured Tom that the Corps supported his efforts to provide information about current restrictions in the area.

Community members told stories about how they hoped communication with the Corps would continue, by stating their intentions to promote open lines of communication with Edwards. As Tom stated, "That'll be an ongoing thing, yeah [communicating with Edwards]. [Edwards] and us. And I think [Edwards] talks to [Frank] a lot, [Frank Byrd, another active community member interested in participating in the preservation efforts], I'm pretty sure." Paul also confirmed the need to continue communicating with Edwards: "We've got a good,

good communication, and a good, uh, plan going forward, I think." Lee voiced a similar perspective:

> So I think, I think it's important to continue to communicate, you know, with a civil tongue, communicate in a way that's, that's, um, considering their [the Corps'] job and their responsibilities, um, and uh, communicate what our desire is for our community, so. I think it's important.

Voicing these stories about future, positive communication with the Corps and others involved in the cleanup efforts set an effective tone for these continuing, negotiated, collaborative efforts. Speaking these narratives of the future could lead to facilitating their materialization and creating a new future reality for all of these participants and voices, a dynamic reality that would be quite different from the conflicting voices of the past, voices that were not communicating aligned values.

Stories About Continuing to Hold Meetings

In addition to stories and narratives about future positive communication with the Corps and community, community members also voiced through narratives the desire to continue to hold meetings. Because these face-to-face, town hall meetings were the site of so much negotiated action between the Corps and community, community members believed these were vital to continued success with these mutual efforts and wanted them to continue. Paul stated,

> And I think that needs to continue [meeting]. Obviously, you need to do both [meet and post information to the general store's Facebook page], uh, that way the word's out. But yeah, the meetings need to be, you know, as long as they give me enough time, I post it [on his Facebook page].

For Paul, continuing to hold face-to-face meetings is part of the process of enacting his own values of making sure all voices are heard and ensuring everyone in the community has access to the same information. Complementing this perspective, Tom anticipated possible changing sensemaking and uncertainty (Walsh & Walker, 2016) at Grey Cliffs based on changing weather and activities, and he wanted to make sure the community "held together" during these changing times:

> I think we need to have, and I think since the summertime, you know, coming up, we might have more meetings. You know, and I think that'll be what will have to happen. To me, I think we should have to have some more because, you know, you're gonna have a lot of people down there that's camping and, you know, hiking, the boats, and stuff, so.

Additional, future meetings would allow opportunities for the community and Corps to generate discussion about the changing dynamics at

Grey Cliffs and potential solutions to any problems that could arise, as a result of these dynamics.

All of these stories and narratives clearly related to establishing a positive future for Grey Cliffs, and all of them included various forms of communication, whether those were stories about collaborating about what should be done with the area, continuing to meet and communicate to negotiate social action, or meeting to discuss any new Grey Cliffs developments. Many of these stories contain antenarrative qualities, since they contain elements of the past and present as they look toward the future (Boje, 2008, pp. 6, 13; Jones et al., 2016, p. 212). These dynamic stories indicated that Edwards, the XYZ Company, community members, and other outsiders using the area had a vision for Grey Cliffs and were invested in long-term growth. Telling these stories of the future characterized these community members as ones who could see beyond the current conflict and completed rejuvenation efforts; this area had a lot more to offer, and the community wanted to help participate in making these dreams more of a reality by talking to people, such as Edwards, who could help make these dreams happen. Stories about changing collaborative actions, changing perceptions of Edwards, and narratives of the future helped develop exactly how these dreams could begin to be accomplished. These dreams were essentially narratives of common value alignment, demonstrated through negotiated actions.

COMMUNITY MEMBERS' TEXTS THAT INDICATED CHANGE

In addition to the narratives and oral stories that promoted community agency and an evolving understanding of community members' participation in the overall Grey Cliffs narrative, the community also used other texts, such as signs in front of the general store advertising meetings, fliers that Norma hand delivered to neighbors and that some community members posted at the general store, newspaper ads also advertising meetings, and social media communication, curated mainly by Paul and Tom. These more conventional texts allowed community members to further negotiate their agency by officially documenting their participation and their involvement in addressing Grey Cliffs' rejuvenation process.

Signs

These signs notified community passersby of upcoming meetings; they were posted next to the road at the general store. Written in chalk on a blackboard or written on white poster board, these notifications included the basics of the meeting topic (Grey Cliffs closure or Grey Cliffs options), date, and time. Only two of these public signs were created by Paul and posted. After that, smaller committees were formed, cell phone numbers were distributed, and later meetings focused on actions to address the conflict that would allow Grey Cliffs to remain open. Paul designed these signs to publicize the meetings as much as possible to community members who would drive by the store and may not have an opportunity to learn about the meetings otherwise.

Fliers

Norma was the primary flier designer; for the first two meetings, she motivated community participation by creating these and distributing them to neighbors through personal visits (see Chapter 5). For cleanup days and other, less public announcements, such as committee meeting announcements, community members relied more on personal text messages and informal, hand-written notes posted to the inside doors of the general store. These types of communication motivated community members to attend the meetings originally, and then community members took ownership of the process of recruiting others to attend these meetings and get involved in the collaborative rejuvenation efforts.

The flier reproduced here indicates the growing support from local-area businesses and details about the cleanup efforts that community members participated in.

FLIER ADVERTISING EARTH DAY COMMUNITY WORK DAY
EARTH DAY WORK DAY AT [GREY CLIFFS]

When: Monday, April 22nd
PLEASE SIGN UP IN ADVANCE ON THE WEBSITE: https://website.register.com
Choose a Shift when you Register:
Shift 1: 8:00 a.m.-12:00 p.m.   Shift 2: 1:00 p.m.-4:00 p.m.

What: Clean-Up of Lake Shore. Description of activity: Volunteers can expect to walk in muddy and rocky areas with dense vegetation to will [sic] remove trash, tires, appliances, & other debris to beautify the community & reduce threats to wildlife.

Closed-toe shoes are **required** (boots are recommended). Wear Long pants tucked into socks, a long-sleeved shirt, & a hat (recommended). Bring a reusable water bottle.

Will be provided:
Work gloves, Litter grabbers, Trash bags, Sun screen, Bug spray, T-shirts, Food & Water ;) [handwritten in]

This Earth Day Event would not be possible without the support of
The US Army Corps of Engineers, The County Clean Commission
The XYZ Company, [Another local company name], [Another local company name] & The Neighbors & Supporters of [Grey Cliffs]!

This document indicates the details organizers used to motivate this community and interested outsiders to participate in these preservation efforts. While this work would necessarily be difficult and involve physical labor, the listing of the organizations on the right-hand side of the flier that supported this activity financially and materially would encourage community members to act. Such community mobilization would then be considered proof that the community could be trusted to keep the area clean, convincing authorities to keep Grey Cliffs open and accessible for public use.

Newspaper Ads

After the first town hall meeting, some community members were concerned that interest in preserving access to Grey Cliffs would falter or that not enough community members were being informed of the meetings. Not being officially connected to a nonprofit organization, though, the community didn't really have a means of paying for additional, more formal advertising. Lee, who was a member of the executive committee, and Trisha, his wife, agreed to draft a newspaper ad and publish it in the local paper to advertise the next meeting. Although the ad appeared, the community did not know for sure how effective it was in motivating additional attendance at the meetings and community involvement. Even so, the effort to draft, pay for, and publish the ad indicated additional movements toward negotiating social action, as well as ethos. After all, the promise that government and local officials would be attending the upcoming meeting, which was communicated in the ad, indicated that the subject matter the community members would be discussing certainly would be important; the ad sanctioned and verified the activity of organizing and publicizing the meeting; this sanctioning in turn helped support the developing ethos of the community as one that could be trusted to align with Corps values.

Social Media Communication

In addition to oral narratives and hard copy documents such as fliers and newspaper ads, some community members were actively involved in social media communication as a way to co-construct an ethos with the Corps, community members, and other outsiders who might visit the area. As discussed in Chapter 5, Norma first created a Facebook page for this effort that required a "secret code" for sanctioned, vetted users to participate. Tom and others rejected Norma's Facebook page because they felt Norma was promoting negative, erroneous information on it.

According to Tom,

> We may have probably done it [created his own Facebook page] because of the way [Norma] had hers, you know, the negative stuff that she had on it really, you know, uh, I think [Frank's] son was the one who brought it up, and I said, you know, we need to do a Facebook page, and I said I'll be glad to do it because we felt like it's social media, we need to get it out there to them, you know, to people that we couldn't reach, that had been down there, other counties and communities, to know that hey, it's not, you know, you heard it was going to be closed down, but it's not, and here's what we're trying to do to keep it from being closed down; you're welcome to help support it.

Tom was trying not only to correct erroneous information but also encourage people to come be involved in the cleanup efforts. Even though this publicity could result in bringing more people to the area who could cause damage, Tom's hope was that through Facebook, he could educate people about what the restrictions were while recruiting additional support at the same time.

Paul also used Facebook to negotiate ethos and agency by posting meeting information on his Facebook page and encouraging community members to attend. He posted information about cleanup days, as well. While published on Facebook along with other events Paul was planning to host at the general store, the meeting advertisements were another way Paul could demonstrate his interest in supporting the collaborative Corps and community efforts to continue the dialogue and cleanup happening in the area.

Importantly, all of these texts communicated the polyphony (Bondi & Yu, 2019) of community voices; while community members shared a common desire to keep the area open, its voices represented individual narratives of the value and worth of this area to each person. Through these texts, the community also responded in symphony (Bondi & Yu, 2019) with Edwards, communicating sincerely a desire to work together to keep Grey Cliffs open. Through a relationship with Edwards, the community took an opportunity to subvert potentially restrictive organizational discourses and change them (Giddens, 1991a, 1991b) to at least begin a negotiation process through these narratives, counterstories, and texts. The hard-copy, oral, and digital texts community members created to reinforce their message of environmental preservation and rejuvenation were key in communicating shared value alignment to anyone who had a relationship with Grey Cliffs, including the Corps. As a result, behind these efforts was a developing trust among everyone involved.

Another important part of the ethos-negotiation process was the ways these changed narratives, stories, and texts translated into actual changed behavior related to Grey Cliffs. Through participant interviews, several community members, including Edwards, indicated how changed behavior supported the changing narratives in tangible ways that continued to build trust between Edwards and the community, indicating more aligned value frames.

COMMUNITY MEMBERS' CHANGED BEHAVIOR: ACCOMPLISHING SOCIAL ACTION THROUGH VALUE ALIGNMENT

Before the town hall meetings began, the areas surrounding Grey Cliffs were turning into mudslides caused in part by off-roading. Erosion due to lost plant life caused deep ruts to appear in the earth leading down to the lake, causing some runoff issues as well, depending on trash and other residue uphill from the lake. Fire pits from camping in the undesignated areas contributed to the lack of plant growth. Beer bottles and cans, general trash, and even used baby diapers were strewn across the area. Hunters and those just looking for a place to target practice had damaged many trees. The visible damage to the area was shocking. These were the environmental problems Grey Cliffs faced in the beginning.

The co-constructed narrative of aligned values for change did result in tangible community efforts to reduce crime, keep the area clean, and eliminate off-road vehicle use at Grey Cliffs. Many social actors were part of this process of following through on changing their actions to protect Grey Cliffs, including Lee and Trisha, Tom and Denise, Paul, Norma, and Edwards himself, as he responded to these community members' changed behaviors and added to them. My inclusion of individual voices in this discussion indicates these participants' importance in experiencing and addressing this conflict; in their own words, they describe the effects that community efforts had in beginning to protect the area. While part of the public impacted by these events, these members were unique individuals who each responded with concrete actions to Edwards's appeals within a community that was determined to keep this area open and usable. The community members were not the only ones who changed, though; these members' changed behavior required recognition and response from Edwards, who in turn supported community members' value alignment efforts with compromises and connections of his own.

Lee and Trisha

As mentioned previously, Lee and his wife, Trisha, owned a small farm that bordered the Corps land that in turn adjoined Grey Cliffs. An old logging road on this farm led down to the Corps land, and this couple was one of two families who often experienced off-roaders trespassing onto their land. Driving up the rough, rocky road, these people would appear literally at this couple's back door. Unable to back down the steep road they had just climbed, they often required assistance with directions on how to return to Grey Cliffs by paved road. Although a less frequent concern, people could also go down to Corps property from Lee and Trisha's land at the top of this logging road, gaining unauthorized access that way for those determined to do so. Lee mentioned that he felt his safety was compromised as a result, and Trisha had always felt that their land was "insecure," based on the inability to keep people from coming onto their property whenever they wanted. In part to help with these trespassing issues, Lee was willing to make some changes to secure his land that also aligned with access restrictions that the Corps wanted. Specifically, Lee stated,

> I was willing to do whatever I needed to do to help keep it [Grey Cliffs] open, even putting gates up on my end, you know, which is an expense out of my pocket, to try to keep it accessible, um, to that, to the main boat ramp area but lock it down on the other end of it [to keep people out], so.

Lee's efforts kept trespassers from leaving the Corps land and accessing his property; putting up the gates confined the activity more specifically to the Grey Cliffs area, which could be policed more easily and minimized trespassing. In addition, Lee indicates that his involvement in maintaining Grey Cliffs as a recovered area will be ongoing:

> Um, I think my role is really going to be more policing the area, um, reporting those who are committing crimes, those who are shooting down there that shouldn't be shooting down there and so on, you know, um, to really stay on top of that and, you know, reporting, [Edwards], he's asked me, if I hear any kind of shooting, to give him an immediate call, and he would come here. So I can see my part is policing the area.

Lee specifically was motivated to install gates on his land to not only protect his property but also keep people on the Corps land. This action also discouraged off-roading and unauthorized shooting, since a large part of the roads used for those purposes could no longer be accessed. Edwards's willingness to communicate with individual community members, such as Lee, demonstrates his knowledge that those closest, most localized, to the impacted environment will have the most accurate information about future incidents occurring there, similar to what Williams and James (2009) acknowledged in their work. The personal connection between Edwards as a Corps official and community members was key to success in making these types of physical changes a reality.

In general, Lee commented on the overall changes happening in the area that the community participated in:

> So with a lot of arguing and fighting to come up with a different plan, [Edwards] decided to keep it open but barricade off one section [of the woods] where all the crime was happening. So, um, so it went from it's definitely going to, the sheriff was definitely going to do all he could to shut it down, to barricade the whole area to the point of now it's going to stay open, and we're going to make it, you know, the community is going to make it nice, you know.

The community overall, based on their love of the area and the desire to continue using it, agreed to be a part of barricading certain sections to restrict the areas for public use. In this case, Lee bought and installed gates himself at two points on his property to help ensure Grey Cliffs could remain open by restricting access to Corps land from his property and also prohibiting off-roaders from leaving the Corps land and accessing other people's property. While the Corps and the XYZ Company provided some of the materials at the actual lake-access point, such as concrete barriers and gravel, the community helped put the barricades and gravel in place to restrict and clean up the area in an organized, team effort. Non-designated-use areas would regrow, and the public would still be able to access the lake and participate in limited, primitive camping.

Tom and Denise

Tom, the local auto shop owner, had grown up in an area near Grey Cliffs and currently lived about two miles from the lake-access point. He had a long history of fishing there, and his wife, Denise, had been baptized there, as Chapter 3 highlighted. The couple had taught their sons to swim at Grey Cliffs. Initially, Tom was very upset about the possibility of closing the area. One of his sons trained for off-road races at Grey Cliffs, which was a convenient location for him; however, Tom couldn't deny the mud slides and erosion issues due to the off-roading. He did ask his son to stop using the area to prepare for racing, and then, he continued using social media to emphasize the changes taking place in the area.

> And I, I took a lot of pictures around [Grey Cliffs], the water's up, and I try to post it, everything, you know, how beautiful it is, and what a beautiful spot it is. But now I went down there and took pictures when they put the

fences up, you know, and I, I tried to write something like uh "Progress," you know. You're just going to be able to camp on one side until they get another side cleaned up. And I said, one side is going to be camping; the other side's going to be for boats and parking. I said we're trying to organize it to where it's a better place for everybody.

On the one hand, Tom took seriously his job of keeping everyone updated on social media about the positive efforts being accomplished at Grey Cliffs to motivate people to visit. On the other hand, Tom was concerned about letting the public know via Facebook that violators of Corps' regulations would, indeed, be prosecuted. Ultimately, he wanted to ensure the success of the community's efforts to minimize the risk of future closure if the area's condition went downhill again. Even though enforcing those regulations was the Corps' responsibility, Tom tried to extend that message to social media as well:

Anybody can go; you know, and we went down there and took pictures, and I even took some close ups of the, the posts and, you know, [the signs about the] four wheelers, and you'll be prosecuted if you pass this point, you know. Even I put on there, not to scare nobody, but you go past this point, you'll be prosecuted. And um, [I] just tell them, you know, hey, this is what's happening, and I'm sorry, you know, you can't take your jeep down there no more, but it's a, it's a public and a family place, it's not an all-terrain area.

Although Edwards had informed community members that they couldn't actually enforce the regulations themselves, due to liability concerns and the fact that the community didn't have the training or legal backing to monitor those not following the regulations, Tom seemed very concerned about letting visitors know the immediate ramifications of not following them:

And like I told [Edwards], if you're going to do that [put up cameras to monitor possible regulation offenders], there shouldn't be no warning. Break the law, prosecute them, 'cause there shouldn't be no warning provided.

Tom's zealousness to prosecute offenders also indicates his limited control over actually enforcing the regulations. While he could control his own actions and encourage his sons not to practice off-road racing there, he couldn't control the actions of others. And if outsiders continued violating the regulations, Grey Cliffs very well could be in danger of being closed again.

Tom continued demonstrating his efforts to align with Corps values by using social media to encourage visitors to throw away trash and not use the area for target practice, which had been other environmental concerns:

> I put on there [Facebook], pick your trash up, please. You do, you know, we expect, you know, everyone to keep their trash picked up. . . . It's actually, it looks real good; last time we was down there, it was clean. And I said there can't be a shooting range down there.

Tom was persuaded to help Edwards accomplish the goals of rejuvenating the area; part of this participation was motivated by Tom's relationship with Edwards. Early on, Edwards singled Tom out as someone who could spread the word in the community, and Tom took that task to heart personally, first by asking his son not to off-road race there, then by going above and beyond by creating and maintaining the area's Facebook page and keeping it updated about the progress made and the regulations being enforced. Viewers of the page could in turn be motivated to participate in these efforts as well, based on real-time updates. Rather than discourage people from visiting the area, Tom participated in cautioning the visitors to be prepared and even help with the sustainability efforts. This type of social media use allowed Tom to present a real-time narrative that served "to offer a sustained argument about ways of being in the world" (Tillery, 2017, p. 52). For Tom, presenting this information was a way for all, including Edwards, to see that everyone's efforts had been paying off and that the previous problems impacting Grey Cliffs were no longer concerns.

Referring to his Facebook page titled "Supporters of [Grey Cliffs]," Tom saw the Facebook social media platform as one that served multiple functions related to changed actions supporting Grey Cliffs: updating the public about current conditions at Grey Cliffs; regulating future behavior by warning visitors of the surveillance, posted regulations, barriers, and reduced activities allowed; and recruiting future volunteers to help with the Grey Cliffs rejuvenation efforts. In addition, Tom's use of social media indicates how it can play a part not only in changing the narrative about Grey Cliffs through text alone, but also connecting that changing narrative, including digital elements of photographs, to physical improvements through changed actions. These efforts were part of a changed, complementary narrative that was a prerequisite for continued, positive change.

In sharp contrast to the restricted Facebook page that Norma had initially created, Tom, on his Facebook page, counterframed the negative messages that would discourage people from visiting the area by citing facts (supplemented with photographs) and Corps regulations that documented what was really going on. Through these social media texts, Tom created a truly dialogic communication environment in which he could inform the public but also answer questions from those who did not live in the area and needed to receive up-to-date information before visiting.

Paul

For Paul, owner of the general store where the town hall meetings took place, evidence of improvement included observations of the committees' functioning, offering drinks and supplies, policing the area, communicating with the Chamber of Commerce, and confirming that Grey Cliffs' conditions had changed by publicizing those changes to customers at the general store.

Observations of Committees' Functioning

While he did not actively participate in the town hall and committee meetings, Paul nevertheless played a big role in improvements at Grey Cliffs, since he owned the general store where the meetings were held. He therefore observed all of the meetings and commented on their success:

> Then we started negotiating on other things, and then we formed, it was our idea to form a community committee. And that we've done. And the committee's been working successful. . . . Now . . . we have a community board and, um, everybody watching [Grey Cliffs], and we have the police now, uh, the sheriff willing to uh patrol it.

Not only did Paul observe the meetings' operating successfully, but he also observed the changes at Grey Cliffs for himself. Having been familiar with this area from long ago ("'cause like I said, I knew the place since I was a kid; my family was there; I knew the place quite well 'cause the [family surname] used to fish all the lands here"), Paul could recall the former beauty of the area as it was being restored and contrasted that with the area's downfall:

> Obviously, it had to be cleaned up, it had to be, you know, run a different way; it had to be managed. Obviously, it was never managed. Uh, and that was the big key of its downfall. Uh, actually, when we started the maintenance, the riff raff lost 60% of it right away. They knew it was the end, uh, and the beautification's already started. Yeah, the beautification has started, you can already tell.

Paul understood the essential role that the committees played in keeping the Grey Cliffs preservation efforts going because of the necessary organizational and administrative roles the community members played on these committees. The community members were strongly motivated by their desires to maintain a personal connection to the area, and Paul supported that desire, too, since his family also had a strong history of positive experiences there. He also networked with officials in Corps and County leadership who visited the store and reached out to the local Chamber of Commerce for assistance ("I've also been in communication with [Sidney] at the Chamber of Commerce"), since Grey Cliffs drew in tourists from surrounding areas. Paul maximized these networking opportunities from his vantage point as general store owner and served as a communication facilitator, also publicizing future town hall meeting times and aid he could provide on the general store's Facebook page.

Offering Drinks and Supplies

In his role as owner of the general store, Paul also offered discounts on food, free coffee, and free water to people involved in policing and cleaning up the area, including officials and community members alike. For Paul, this was something he could easily provide and rationalize:

> But I did offer them [Corps and County officials] free coffee, free water. They came in and ate lunch, whatever, and I gave them a substantial discount, all part of my contribution. But I give kids ice cream; I don't forget where I came from. What's a scoop of ice cream? Like water. The hikers come in here and the bikers; I give them water. 'Cause I've been places, asking when I would have given a hundred dollars for that bottle. I mean I get it for four dollars a case. You know, I sell one or two? The rest of it, I give it to the people that, hikers, you never know? And people freak out on that; I'm like, why would you freak out on it? It's water? 'Course I need to keep the lights on, but other than that, I don't care.

Paul saw these provisions as a way to encourage the monitoring and cleanup efforts, perhaps also as a reward for working to preserve the area:

> Yeah, and I was posting [on Facebook], you know, like once we started getting it [the cleanup efforts] going, you know, I had people come here and get free coffee, and you know, whatever I can do to communicate to have them continue down there.

In addition to discounted food, free drinks, and publicizing this support on Facebook, Paul also had some ideas about other things he could donate to help make monitoring the area a bit easier and more comfortable for those involved. For example, Paul owned a cabin that he offered to donate:

> I've got a cabin up there; I was going to contribute it to them, uh, to use as a cabin or police to stay in or whatever, for them to get out of the weather and whatever; it's up on the hill. That was part of my donation.

The cabin idea extended also to enhancing primitive camping in the area, which was the only kind of camping allowed during the rejuvenation efforts:

Well, I mean, if they need something, like if they plan on keeping uh primitive camping down there, whatever, I may could have more cabins, you know, uh, you know that they can use or whatever, but uh, but yeah, just, whatever they need, basically that I can do within my power.

Paul discussed these donations as an ongoing activity, indicating his long-term support of these efforts, which he assumed would be taking place over a long period of time.

Policing the Area
Although Paul was "married to this business" of managing the general store and could not spend a lot of time relaxing at Grey Cliffs, he did indicate that when he had time, he would participate in policing the area, too, along with other community members:

Well, yeah; my role will be to help to support them [officials and community members], uh, in every way that we can. And uh, you know, when I got some free time, go down there, and if we see anything out of the ordinary, we report it. Uh, yeah, it's, we constantly got to keep an eye on it. And we got to support whatever they need. Whatever is within our means.

Along with other community members, Paul emphasized his willingness to help out in whatever ways that he could; he acknowledged his limitations regarding time and needing to "keep the lights on," but he found ways to support the community and their efforts to comply with Corps regulations as best as he could.

Confirming Changes by Talking With Customers
Paul's position as general store owner and manager placed him in the unique role of being able to recruit visitors to Grey Cliffs, which was located about two miles from the store, and to keep people posted about the area's status. Some people would mention to him that they had heard the area was closed, and Paul updated them with correct information and also the progress of the cleanup efforts:

We've had customers in here today went down there riding [motorcycles], from [a nearby city]. And stopped in here and ate. You know, and then talk about, uh, man it really looks good, you know, I mean it's a good compliment for our community.

Paul could affirm these visitors' observations and experiences, no doubt extending the conversation with details about everything the community was doing to enhance the area.

During our interview, I observed out loud that Paul seemed to be a key part in helping keep Grey Cliffs open, based on hosting the meetings, providing food and drinks, and offering to help out further. His central role as in-person and social media communicator via Facebook also extended his outreach to basically every segment of the population. In response, Paul confirmed, "I would like to hope so, and that was my goal." When asked about his opinion about the overall progress taking place and whether he felt good about it, Paul also confirmed, "Yes, most definitely." For Paul, as well as for other community members, seeing such immediate and significant results from their changed behaviors seemed to provide more motivation to continue this work, even though it was challenging, time consuming, and somewhat costly in terms of donations.

Norma

Yet another community member who negotiated agency to some degree and responded to Edwards's ethos appeals of experience, expertise, sincerity, and affinity was Norma. Norma's efforts did not go so smoothly—at least not from her perspective—even though Edwards had initially persuaded her to act. As outlined in Chapter 5, after speaking with Edwards about changes needed at Grey Cliffs, Norma began the grassroots effort to organize the first town hall meeting. She canvassed the neighborhood on foot and distributed fliers about the meeting, continued communicating with Edwards, and emailed and made "phone calls to other county and state officials." She "promot[ed] [a] public forum [she] organized (acting as liaison between the Corps, county officials, other interested parties and 'neighbors') thru face-to-face canvassing of neighborhood and social media." Norma was very pleased at the early, positive result and confirmation of her efforts. Being persuaded by Edwards that something needed to be done to help the area, Norma chose the solution of creating a nonprofit entity that would be the overarching framework for these efforts, which were crucial in jump-starting this community to position itself to act. Norma had additional goals in mind beyond what the Corps wanted, though, and those goals were not as easily accomplished as what Lee, Tom, and Paul had done through their work on installing gates, conducting surveillance in the area, and keeping the community updated on progress and regulatory efforts through social media.

EDWARDS'S SUPPORT OF A DEVELOPING NARRATIVE: ACKNOWLEDGMENT OF PROGRESS MADE

This discussion of the community's changed behavior would be incomplete without Edwards's reaction to these changed narratives, texts, actions, and plans; he would need to accept the revised, negotiated ethos of the community and be willing to continue interacting with it, as well, in order for Grey Cliffs to remain open. Edwards had originally (and rightfully) framed the issues at Grey Cliffs as very serious ones that merited action such as closure, and the community had initially presented a very strong counterframe of resistance, community ownership, and dedication to the area, including its own antenarrative and counterstories. Based on this resistance, Edwards began negotiating with the community in a way that would provide room for these initially opposing groups to co-construct a counterframe with him that promoted a narrative of change. In the process of co-constructing this counterframe, Edwards reflected back community efforts in a positive, new narrative of a revised Grey Cliffs identity from a Corps perspective. This identity included value themes such as compliance, new community relationships, community buy-in, corporate buy-in, and framing a positive future.

Compliance

From the beginning, the most noticeable problem at Grey Cliffs was that the community, both insiders and outsiders, were not complying with the rules and regulations (essentially the laws) that were set forth by the Corps as a government organization. However, Edwards had acknowledged that, when attempting to move people toward compliance, they may not comply long-term because "rules and regulations are black and white"; the relationship-motivating connection was absent. Fostering genuine relationships ensures compliance better than emphasizing rules and regulations alone. Through fostering relationships with the community, Edwards did see the compliance he was looking for:

> And so, through these meetings and through word of mouth, and through the signs we've published that people now know they can't ride their ORVs there. And since then, I have seen, I haven't seen a single use of uh ORV or ATV in the non-desig in the area that we designated illegal activity, you know, with what you all are doing down there . . . , and up the creek there, I mean I've walked it several times, and there's been no use. Just foot traffic. It's been incredible. I've seen very little if any activity other than hiking and mountain biking in the area that we've designated for, uh, no recreation, no motorized vehicle activity.

Through video surveillance and personal observation, Edwards compiled a well-rounded picture of what exactly was going on at Grey Cliffs. The physical results Edwards saw were in full compliance with Corps' rules and regulations.

New Community Relationships

Persuading this community to comply with Corps' regulations coincided with Edwards's developing relationships with community members. In his interview, Edwards admitted that in the past, he had not done much personally to connect with the community because the community seemingly did not support the Corps:

> Anything that I'd done previously—I'd put a port-o-jon there; it was burned to the ground. I put signs up; they were shot. Um, you know, we, we put signs up, no ATVs; they were removed; they were shot.

To Edwards, there was obviously strong opposition to Corps management of the area, and Edwards could have interpreted these actions as threatening his authority. However, during the town hall meetings, Edwards took the opportunity to

> build great relationships by showing that I was invested in the future of [Grey Cliffs], and they asked some tough questions and specifically, you know, how much money had I spent, what had the Corps done up until that point to change the dynamic. . . . I think a lot of the locals who were contributing to the problem admittedly, um, said they would stop, and they would ensure that, you know, whoever they knew that were doing, they would also stop. And since then, the whole dynamic of [Grey Cliffs] has shifted. The paradigm has shifted. The support of the community has made to me the difference.

Edwards needed to demonstrate his interest in and investment in Grey Cliffs and the surrounding community. Part of this process was answering some difficult questions from community members. A related issue Edwards had to address was the apparent lack of Corps presence in the area because the evidence of Corps presence (in the form of signs, port-o-jons, etc.) had been removed. Edwards stated that community members had said to him,

> "We never see you all down there; we never see the government down there. Who manages the area?" Well, that was to me identified as, well that's an issue; they need to know who, who's responsible for operating and maintaining the area. So I kind of noted that that was something that I needed to improve on.

One thing Edwards later did to improve Corps visibility was post official signs with the Corps logo, colors, and clear labeling so that anyone visiting the area could clearly see the relationship between the Corps and Grey Cliffs.

In being honest about his own accountability and responsibility to the area, Edwards also encouraged community members to be responsible for their actions. Edwards demonstrated this principle at the first town hall meeting by emphasizing that he liked to go off-roading, fishing, and camping, as well, but he always asked permission from landowners and made sure to follow the rules. He attempted to lead by example in this way and emphasized during his interview the need for that continued responsibility:

> So I attribute the personal connection that I made with them and um the fact that they knew that the um you know what they were doing was illegal, and it was causing damage, you can, I mean, you don't have to look far to see the damage and destruction that's been done through the ORVing and ATVs there, and I'm glad they, you know, were, with them there, I think it really carried a lot of weight with their support of the program or the proposal to keep it open and keep the ORVs off made a big difference, too.

Part of Edwards's process for creating this revised, shared narrative was recruiting ambassadors from the community, people who had participated in the off-roading and acknowledged it needed to stop. Those community members could then spread the word to those not at the meeting and encourage their participation in these positive efforts, as well. The fact that Edwards, the compliant community members, and other community members who needed to stop these activities all had common interests, an affinity for the same types of activities, helped this message of compliance be more palatable, persuasive, and feasible to accomplish.

Many kinds of affinity could have played a part in this relationship building. Affinity may have grown, based on participants' investments in a common, rhetorical situation within an evolving discourse community (Paré, 2014). All of these social actors, linked through a common interest in a geographic space and physical location or "affinity space" (Gee, 2004, 2005; Neely & Marone, 2016), also communicated within "interest-driven social environments in which participants engage in passionate, self-directed, and intrinsically motivating activities" (Neely & Marone, 2016, p. 59). These common activities also contribute to a bonding process. Importantly, this type of affinity is strongly related to "*spaces* in which such interactions occur and the shared *interests and passions* that propel them" (Neely & Marone, 2016, p. 60). Common interests related to space, therefore, necessarily unite affinity groups, rather than common, participant demographics alone. Even so, affinity between Edwards and the Grey Cliffs' community could also have grown based on the appearance of a common ethnicity and race. While this topic did not reveal itself at all in the data, on the surface, all participants seemed to have a relatively similar ethnicity. When Edwards spoke, too, his words most likely resonated with community members, as they were communicated using a similar, Southern accent. While no conversations took place about where Edwards was from, the growing affinity between himself and the audience could have benefitted from this oral connection; community members may have accepted Edwards's persona more, even subconsciously, as a result of these perceived similarities, and these physical affinity characteristics could have facilitated the personal relationship development between him and the community, in addition to the growing participation of all social actors within this affinity group.

After the initial upset at the first meeting, Edwards not only reached out to people at the meeting to try to build relationships and increase this affinity, but he also continued afterward by talking with community members by phone who were concerned about possible closure:

> I had a couple of phone calls and some conversations with some individuals that called, and I met with [them] on site after [the] meetings, and, um, you know, further expanded on, you know, what we were trying to do, and I think they got, I think they finally understood that yeah, okay, there's an option here, that we're not willing to close it, we're wanting to keep it open, but what's the, you know, alternative.

Edwards could have begun by simply stating at the meeting that other alternatives existed besides closure; however, the community didn't seem to understand all of the implications. In an effort to change the Grey Cliffs narrative, Edwards took the time to explain these different options to interested community members eventually and explain the Corps rationale for these efforts, in an attempt to develop relationships with these community members.

Far from these new relationships' being one-sided, instigated by Edwards alone, Edwards conveyed a reciprocal vision for these interactions:

> The public meetings really produced a lot of, um, I think it brought the community and the Corps together and gave us the opportunity to have some credibility, uh, in the area because we were making the investment both, you know, from a financial, fiscal perspective, as well as, respecting, you know, what the community's wishes were.

Edwards hoped that, in his role as Corps resource manager, he could continue demonstrating sincerity and credibility through a financial investment while at the same time supporting community wishes to keep Grey Cliffs open. The community ideally would respond, in turn, with its own volunteer investment of time and effort to continue supporting Corps regulations. This relationship, while in a sense a business one, especially from the Corps' perspective, also aligned with very personal community values, all revolving around the use of Grey Cliffs.

Part of the new narrative for Edwards included not only a sincere, trustworthy, and credible persona as indicated through his own words during the town hall meetings and interviews, but he wanted to present the community with a new vision for what Grey Cliffs was supposed to look like via his mission as Corps resource manager: clean and maintained. The character traits Edwards endeavored to display closely related to the Corps mission he represented and the desired action he hoped to continue seeing from the community. Edwards believed that if the community could actually see this vision realized as he continued interacting with community members through these relationships, they would be more invested, too, in maintaining it, as he was.

Community Buy-In

A successful, changed Grey Cliffs narrative included relationships Edwards had attempted to build to co-construct that narrative, but it also included community buy-in from not just these individuals but other community members. Edwards saw evidence of this buy-in during cleanup days: "And the buy in specifically, I mean, we've had what, three, four, uh, what I call work days at [Grey Cliffs] where we've put up the barricade, we've spread gravel, we've picked up trash, we have, you know, installed signs." These efforts demonstrated significantly the changed narrative that the community had adopted/adapted; previously, visitors to the area had destroyed the signs, gone beyond the barricades, and left trash behind. Now, community members were actually volunteering their time to contribute to the cleanup efforts.

More invested community buy-in involved organizing committee meetings, which was an administrative outcome that Edwards never expected:

> They [community members] have met individually and formed, uh, you know, an executive committee, you know, coming up with ways that they can help be a part of the solution and not the problem. And, and that is to me, I mean, that [first town hall] meeting, had I known that this would come from that, I would have never guessed that in a million years. You know, I would not have thought that that would be a part of the outcome because of, going in, I knew that, you know, [Grey Cliffs] was just one of those areas that people go to, you know, get out of the eyes of the law and kind of raise cane.

These meetings led to the organized improvements that community members participated in; Edwards realized and acknowledged the role that community values played in the continued motivation to invest in these improvement efforts:

And because the community, you know, it cared so much about it, and there was so much history there, and so much cultural, uh, attachment to the area, they, you know, generationally, their fam- their granddads had come there, their, you know, they've taken their grandkids there, I mean it's one of the only, you know, great accesses [to the lake] in [the] County. They were very passionate about keeping the area open and doing everything we can to, to change [the conditions there]. And so we came up with the idea to barricade the area, we barricaded it. So yeah, they [community members who were off-roading] have spread word; they use their pages to spread word, and they know, they knew the community was doing it, I mean the local community that were off-roading there, and they stopped it.

Edwards was also surprised by the length of time that the Corps and community efforts were continuing. In the past, Corps signs and barriers had been removed no sooner than they had been put up. Although Edwards could not tell for sure who had removed the signage before, a difference now seemed to be that the local community members who were invested in the area and had actually installed the signs were dedicated to these continuing improvements, as a result.

But more of the locals that were more invested, uh, in the community, and I say that meaning, you know, they're from there, they grew up there, they moved there, they have, you know, raised their kids there, etc. You know, like, your family and, um, you know that, those people have more of a personal investment like I do. . . . And um, people started proposing ideas, you know; there were a lot of suggestions made, and through the executive committee, the couple of meetings we had the barrier idea was proposed, and I never would have thought the barriers would have lasted this long. I expected them to be cut; I expected them to be pulled out; I expected them to be driven through, and they haven't been touched. And I only say that because, I suspect one that they know that we're watching the area, you know, we are doing surveillance there, and we'll prosecute if they are damaged, but I think because the locals, you know, they invested sweat equity into it; I mean, they stained those barriers; they installed that cable, and so when they're down there, they're the ones that when the public says, "Why did they put this fence in?" The guy says, "Well, I did it!"

Edwards envisioned an area where community members would be talking about the area and would agree that they did not want the old, eroded, environmentally damaged landscape that had become Grey Cliffs. Edwards admitted that to some degree, "they've done the work for me, up until a point. Um, but I think it's built on the trust that we're going to generally make a difference and do something that has a long-term, positive impact to the future use of [Grey Cliffs]."

For Edwards, community buy-in reached beyond community members' physical efforts to communication efforts such as "word of mouth," conveyed in part through social media channels that other community members such as Tom had already attested to. Edwards confirmed:

> Um, and again, the community that's been down there has, we're doing the work, and the word of mouth that they have used, you know, that word of mouth, the weight that [those messages] carry. A couple of gentlemen down there [in the Grey Cliffs area] that, very vocal, and, um, you know they've been using social media for their, their own personal social media webpages, they've used, um, a public page that they've put out there to educate the public about things and events that are going on, and so they're directing questions and answering questions that they get on their pages and directing them to this public page, supporters of [the] County I think is one of them; there's a couple of them. And so they're keeping the public informed about what's going on; they've posted the pictures of the signs there that we've put up so the public's being educated about the rules and regulations.

Importantly, the general public needed to have access to this changed Grey Cliffs narrative. Yes, most community members and the Corps had bought into it, but visitors to the area came from places that had no knowledge of this newly constructed narrative. The social media presence that Edwards noticed and acknowledged played a critical part in expanding the community buy-in to visitors who were not officially part of the community but yet would still play a role in constructing that narrative, for better or worse, through their briefer interactions with the area. These more fleeting interactions were just as important to maintaining the area as the more consistent, intentional ones by local community members. This type of public participation and recruitment to help address Grey Cliffs' problems is similar to other community efforts that have addressed environmental concerns, such as those discussed by Williams and James (2009); these scholars discuss ways environmental advocates and minority groups can participate and have a voice in environmental regulatory actions.

Corporate Buy-In

In order for these efforts to be successful long term, Edwards also saw corporate involvement as a necessary part of support, such as positive publicity and funding opportunities. Many corporations, such as the XYZ Company, invest in charitable efforts such as this one through their nonprofit connections and roles, and the XYZ Company was no different. These companies see contributing to environmental causes as part of their corporate sustainability mission, and, ideally, the corporation would also partner with impacted local communities to ensure that these continued efforts are beneficial to communities and also involve community members themselves. As an example of this type of partnership, when the initial conflict began and Edwards determined the community's interest, he, Tom, and Dan planned to visit the XYZ Company to determine if some sort of alliance could be made, one that could possibly enhance the XYZ Company's mission and image while also furthering the goals of rejuvenating Grey Cliffs at the same time. While Tom had reflected on the meeting he participated in after the fact, Edwards's interview took place before the meeting with Tom and Dan, and he discussed his hopes for this particular meeting during our conversation as an indication of continued success in making positive change:

> I got a meeting with [the XYZ Company] tomorrow; this is the third meeting that we'll have, and they are interested in a long-term, 10-year, they want to form a 10-year memorandum of agreement where responsibilities are lined out and a phased approach towards making improvements to benefit the public in terms of recreation and environmental stewardship activities. And so these are, you know, realistic, um, initiatives that are going to bring long-term, long stability to the area that makes it sustainable on a very low, you know, um, operations and maintenance cost. So in other words, the money that the federal government spends is lessened because of the partnership that has come from this, these meetings and this, this hardship that the public was put through by the recommendation to close it. I mean the list goes on.

This co-constructed narrative included Corps/government, XYZ Company/corporate, and community/group participants, as well as individual social actors within all three of these categories. While all had different motivations for participating in these efforts, all parties demonstrated evidence of framing a more positive future for Grey Cliffs, since everyone was dedicated to participating in these efforts for the long term.

Framing a Positive Future

For Edwards, another part of verifying the positive results taking place in the area was envisioning applying this success to other cases. He stated,

> This is to me a model of, you know, what can happen. And that's what I hope it turns into in that we have people who have made an investment of, you know, personally, professionally, through, they're taking time off to come down and help for work days and projects that we're doing, and we fully intend to make the area a safe and sustainable place for the public to enjoy [the] Lake. Um, this is not anywhere near over. . . .

To date, the Grey Cliffs efforts had been so successful that they became a model of future action. This action would continue to involve other government organizations in the region:

> In fact, the cleanup that we have scheduled for April is going to be a partnership between the [name of county] County Alliance Commission, the Clean Commission, the . . . County Mayor's Office, they're going to bring the dumpsters down.

This positive discussion—framing the future, positive collaborations among government organizations, communities, and more lake-access points—verified Edwards's continuing affirmation of the work already done and possibilities for future applications of that success.

Looking toward the future also suggested areas that could still use improvement, though. Edwards was fully aware that, because the original conflict and the resolution's beginning had taken place during the winter months, only later, during future summer months, for example, would the boundaries and limits be tested enough to determine long-term success. Edwards admitted,

> I would say it's [the situation is] evolving, I would say that it has a bright future, I see an area that kids are going to be able to go to and learn from and experience, you know, the, uh, great outdoors, and there are several initiatives that we can tie into the community.

Characterizing the narrative as evolving also provided room to accept those community members who were still not fully on board with these plans and efforts.

> [There are] [a] lot of people monitoring, you know, observing what we're doing, and I still think those are the outliers that haven't fully invest[ed], you know, bought into, that hey the government is here to help in this instance. They [government officials] want to do good.

As a government representative invested in sincerity about the future as well as the present, Edwards emphasizes his own role and responsibility in addressing these long-term efforts:

> And so, you know, my actions, the actions that we'll have, at [Grey Cliffs] will demonstrate, continue to demonstrate that we're sincere and that what we've done to this point is not going to just go away and be wasted; it's going to make I hope to have a long-term, major impact to the overall dynamic and mindset of the visitor to [Grey Cliffs].

Clearly, the "we" that Edwards refers to includes the community members and the XYZ Company, for example, that are also invested in this future narrative and vision based on current success. The narrative has changed from a place where people can go to "raise cane" to "it's no longer a place they can go to get out of the eyes of the law," according to Edwards. Additionally, this evolving narrative included the following discussion of additional financial resources to increase patrols in the area to minimize crime:

I've just increased my law portion of the contract by oh, what did it go to, I think it was $15,000 more, or so; I think it went to $29,000 this year, and all those additional patrols are scheduled for [Grey Cliffs]. Um, I've met with the District [X] Judicial Drug Task Force, you know, there's going to be some undercover, you know, type activity there.

This type of significant effort to move this narrative forward contrasts strikingly with previous stories I heard from community members living in the area that stated law enforcement was rarely present at Grey Cliffs, including the Corps itself. In contrast, now, according to Edwards,

> Monitoring Facebook, monitoring [Grey Cliffs] messages, you know, I no longer see what the, what's the Corps doing, they're fools, they don't know anything about [Grey Cliffs]. That changed into, "This is great; look at the work they've done," you know, we've put the gravel. . . . And that's what I hope it turns into in that we have people who have made a[n] investment of personally, professionally, through, they're taking time off to come down and help for work days and projects that we're doing, and we fully intend to make the area a safe and sustainable place for the public to enjoy [the] Lake.

The new framing of this narrative highlighted engagement at all levels from the Corps to the community to local law enforcement, including physical and financial resources that would help make it all happen. This narrative could only have moved forward with the affirmation of Edwards as resource manager, who was a witness to the possibilities for change in the area now that he had participated in it firsthand and had recruited the community to help. What began as a narrative of crime and discontent was reframed as a narrative of possibilities, then to affirmation, and then to future plans for stability and continued success, all the while including acknowledgment that community support efforts could continue to improve along the way.

CO-CONSTRUCTED NARRATIVES AND VALUE FRAMES, AN ALIGNED ETHOS

A crucial part of resolving this conflict required community members to develop an anternarrative and counterstories that would not only contrast with the criminal narrative and stories Edwards had originally painted of them but would also invent and anticipate the community's successful role in preserving Grey Cliffs. Community members counterframed Edwards's characterization of them as not criminals but people who sincerely cared about the area and wanted to help preserve it in any way that they could. During this negotiation process, both Edwards as an individual and the community members tried out different frames and counterframes until both parties constructed a narrative, stories, and ethos that both parties as a whole could accept and work within, based on aligned values for now.

Co-Constructed Narratives

Similar to their apparent lack of cultural capital, the community, at first, appeared not to have much agency at all, based on their narratives, lack of compliance, positionality, and apparent lack of "official" authority to enact change. Opportunities for negotiating agency seemed black-and-white at first: either the community obeyed the rules, or they would not be allowed to access this area. Similar to how Edwards ultimately realized his need to communicate using a different narrative that included relationships beyond the black and white of regulatory communication, community members recognized their need to negotiate the terms of compliance to ensure a continued connection to this geographic space; the community ultimately had to realize that the old narrative was no longer working; they needed to create a new narrative, one the community could collaboratively negotiate and own as part of a new story. Most importantly, the community would need to demonstrate compliance with the rules and regulations in order to participate in any compromises. While the community did not understand at first how they could feasibly negotiate with Edwards as a Corps representative, with Edwards's help, they did eventually understand their part in accepting the Corps' rules and regulations as genres of compliance.

As a result of the community's motivation to negotiate with Edwards to keep Grey Cliffs open, the community changed their narrative from focusing on Grey Cliffs' virtues alone to a narrative of these virtues aligned with the rules and regulations. In other words, the community's negotiated new narrative gained acceptance not only for the community but also for Edwards and even broader audiences, such as the XYZ Company. The XYZ Company was not going to pledge financial support unless it accepted this revised narrative of change, as well. This narrative could not be one that was imposed from above by Edwards or anyone else; it needed to be co-constructed and originate from all parties involved. This "rearticulation of cultural rhetoric" (Herndl & Licona, 2007, p. 150), such as what community members and Edwards participated in, is what Herndl and Licona (2007) characterize, ultimately, as agency (p. 150).

This example of co-constructing a new narrative to accomplish joint social action exemplifies Giddens's structuration theory (1984, 1991a,

1991b) in practice: yes, certain power structures and genres need to be internalized, recognized, and practiced in order to accomplish social action ("As organizing structures, genres shape beliefs and actions" [Yates & Orlikowski, 2007, p. 69]), but these structures and genres do not necessarily *determine* "how organizational members engage in communication" (Yates & Orlikowski, 2007, p. 69). This "enabl[ing] and constrain[ing]" (Yates & Orlikowski, 2007, p. 69) process clarifies that social agents involved in negotiating agency and social action are not completely free to accomplish action, nor are they completely constrained (Walton et al., 2019); there is a constant give and take, a cyclical and evolving process that takes place in fluid, unpredictable ways, as the negotiating parties and circumstances themselves change during their social interactions. This type of give and take means that one side doesn't necessarily prevail as individuals and groups negotiate communication related to changing technical, scientific (Wynn & Walsh, 2013), and environmental (Tillery, 2019) goals. This process also includes incorporating marginalized perspectives and participation, a requirement for socially just and inclusive change (Walton et al., 2019).

Significantly, changing beliefs are part of this process: to some extent, the community had to believe that complying with the rules and regulations was part of the solution of continuing to have access to Grey Cliffs. These beliefs were also reflected in how the conversations and narratives changed about Grey Cliffs into ones that would be more acceptable, based on the ultimate goals for the area. No longer would off-roaders tell stories about their latest races or runs in the area. Instead of these stories, community members would share knowledge about other places to off-road, or their activity would have to cease for a while. Community members would share contrasting stories, such as Tom's, which focused on encouraging family members not to use the area, participating in the cleanup efforts, and communicating to broad audiences via social media. Because of the social nature of co-constructing agency and, therefore, power, framing this interaction with co-constructed frames became essential: these newly negotiated frames would reflect also negotiated values, cornerstones for the potential collaborative success of these efforts, as well as an aligned ethos.

Co-Constructed Value Frames

The process of framing and counter-framing often involves contrasting narratives and a resulting frame that represents a more accepted view of a contested situation, especially from the public's point of view (such as what happened with the SeaWorld controversy that Waller and

Iluzada [2020] discussed). The Grey Cliffs community and Edwards took their contrasting counterframes, narratives, and stories and together co-constructed a frame that is stabilized for now, based on aligned values and a resulting aligned ethos. Framing, counterframing, and co-constructed frames change constantly, based on changing social structures, communication, and even outside influences such as laws and policies. The Grey Cliffs' co-constructed frame, therefore, is just a snapshot in time of ways Edwards and the community started with values that did not coincide well but compromised and agreed upon a few values that were central to the conflict.

Part of this process was the community's counterframing the closure option that Edwards presented with alternate solutions, communicated through antenarrative and counterstory. Edwards, based on his authority, helped by countering back with parameters for these revised suggestions for improving Grey Cliffs and ways they might work. The community, due to their lack of experience, had no idea how feasible these options might be, but Edwards's co-construction of these solutions, based on his experience revitalizing other lake-access areas, opened the community's eyes to possibilities that had potential.

Figure 7.1 presents Edwards's initial values as well as the community's initial values, based on the discussions at the town hall meetings as well as the interview transcripts. Evolving, aligning values appear under the initial values for Edwards and the community, followed by a set of aligned values that everyone could agree on. Not all values were candidates for alignment based on the unique, individualized nature of some of the values. For example, Edwards, as a Corps representative, made special efforts, as discussed in Chapter 6, to present a sincere and trustworthy persona. He was also motivated by the Corps' vision and mission to communicate ethically with the public. Similarly, based on Corps' values, he committed himself to furthering the ultimate goal of sustainability regarding the use of public lands and building trust and relationship with the community. The community originally did not identify with any of these values; many of these were ones that related to Edwards individually and his purposes as a rhetor and were not directly connected to the community. The community counterframed what Edwards was saying about values with their own stories of ownership, stories that reflected the values that were important to them: religion, tradition, recreation, skepticism of outside involvement, and unity among community members.

As Edwards and the community interacted and negotiated over time, these values began adapting and changing, based on values that potentially *could* align, to yield a common set of values that aligned during

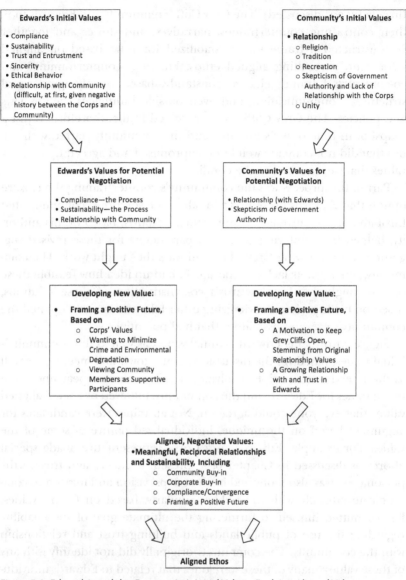

Figure 7.1. Edwards's and the Community's Initial Values, Evolving Aligned Values

the negotiation process. During this process, too, another new, jointly constructed value emerged: framing a positive future, which was based on different values and motivations from both sides.

Similar to how the community had difficulty identifying with Edwards's values, at first, Edwards did not identify with or simply was not aware of the community's initial values of religion, tradition, and skepticism of government authority, since he was not from the Grey Cliffs area. He could not share in the individual stories of community baptisms or the recreational narrative of hosting family reunions, picnics, and swimming lessons there. While distrusting governmental authority was not a value Edwards held himself, he did indicate through his interview that he understood "anti-government sentiment" and its origins; he kept that awareness in mind when preparing to speak to the community at the first town hall meeting. This skepticism was probably the only community value Edwards was aware of when going into this conflict. The community, although it should have been aware of Edwards's values of compliance, sustainability, and ethical behavior, especially, did not seem knowledgeable of these values based on their violent opposition to Edwards initially at the first meeting.

However, some of these initial values from both sides presented more potential for alignment, and those became the ones that helped promote an agreed-upon resolution between Edwards and the community. Of Edwards's initial values, the ones that presented the most potential for alignment with the community were compliance, sustainability, and relationship with the community. Compliance as a value was nonnegotiable. Even though Edwards saw some flexibility in how to communicate it and motivate the community to participate in it, he also saw the problem of relying on rules and regulations alone. While the rules and regulations themselves were black-and-white, Edwards's own experience indicated that this particular community had not responded well to those alone; he had evidence of the posted regulations at Grey Cliffs being destroyed and forcibly removed. While Edwards certainly could have enforced the rules and regulations regardless of their physical presence and regardless of opportunities for relationship with the community, he looked at the "big picture" view of his role and the relationship among the Corps, Grey Cliffs, and the community, and he decided to allow the *process* of compliance to become more negotiable. In essence, he was willing to accept the idea of convergence. While the end result would be the same, ultimately, the community involvement allowed the process to become one encouraged by relationship rather than the black-and-white binary of the rules and regulations themselves. This type of convergence process indicates the cyclical potential for continued negotiation and alignment, particularly for environmental risk communication:

> Risk communication couldn't just be a fast broadcast; it also had to be a slow dialogue between expert and nonexpert communities. And it couldn't be one-off; it needed to be iterative, until these communities could converge on a common solution based on shared values. It couldn't

be linear; it needed to be cyclical in order to respond to cascading global risks. (Olman & DeVasto, 2020, p. 18)

Compliance, tempered by convergence, therefore, became a value that the Corps and community could agree upon as an aligned value, and it was based on the collaboration between the "expert" (Edwards) and "nonexperts" (community members). This focus on convergence helps ensure longer lasting, positive results than a focus on total compliance or conformity would, since compliance can often lead to complacency, which could in turn lead to a reoccurrence of undesired behaviors (Sauer, 2010).

Edwards's value of sustainability was one that the Corps promoted: the Corps' responsibility was to maintain these public lands so that they could continue to be used by current and future generations of the public. Eventually, once the Grey Cliffs community was able to look beyond the crime statistics they did not feel responsible for or have control over, they agreed that something had to be done to repair the damage from off-roading and the impact of camping beyond the intended Corps boundaries. Other community members indicated agreement through their dedication to cleanup efforts (Tom, Dan, and Paul, for example), informing the public about the changed regulations (Tom and Paul), following the regulations themselves (all community members closely involved with the conflict resolution), and contributing their own funds to restrict the area geographically (Lee).

Relationship also became a value that Edwards found in common with the community. Of all of the community's initial values, most of them entailed some form of relationship: religion, tradition, recreation, and unity. These values required relationships with others; religion and tradition especially required the participation of others who shared similar beliefs, and the value of unity did too. Even the final community value, skepticism of government authority, in a way related to relationship because the community strengthened its insider status by aligning itself against the government and Corps outsiders. Because this close-knit community already valued relationship so much, the challenge was accepting Edwards as a friend, at minimum a trusted other the community felt comfortable collaborating with. Based on Edwards's efforts at building his character and experience and demonstrating those in a trustworthy, sincere manner that was supported by his actions, the community members finally accepted Edwards as someone who could help them resolve this conflict, someone they felt they could establish a meaningful relationship with. Sustainability as a value was also closely connected to these relationships since all of the aligned values entailed relationships in some form.

During this process of changing and aligning values, a new value began developing for both the Corps and community: framing a positive future. For Edwards, a positive future included foundational Corps values, wanting to minimize crime and environmental degradation, and viewing community members as supportive participants. From the community's perspective, framing a positive future was based on a motivation to keep Grey Cliffs open, grounded in the community's original relationship values. Because relationship was so important to this community, a growing relationship with and trust in Edwards was a significant part of framing a positive future. This new value, although originating from different Corps and community perspectives, depended on positive relationships between these two groups, as they focused on framing a positive future for Grey Cliffs.

In addition to meaningful, reciprocal relationships and sustainability, community buy-in was another aligned value that both Edwards and the community could agree upon. Edwards knew from past experience that he could not solve this problem alone. Yes, the regulations could be forcibly enforced, but one of the Corps processes was to involve the public in discussions of change to the access of publicly used lands. The Corps would inevitably need to interact with the public during this process, and ideally, they would be able to work together to accomplish this change. Likewise, the community ideally would rely on many members to help monitor the area, keep it clean, organize cleanup days, maintain it (sometimes with their own equipment and resources), and communicate about any changes being made. As the conversation about this conflict continued moving forward, all could agree that community buy-in as a whole was necessary for this effort to be successful long term.

Because all could agree that this process would be ongoing and that many different types of resources would be needed to sustain it, both Edwards and the community also agreed that corporate buy-in was necessary as well. While the Corps had some funds available for improvements, ideally corporate buy-in would introduce other levels of financial and resource support than just one governmental organization could provide initially. This aligned value demonstrated its importance during the joint meeting between the XYZ Company and Edwards, Tom, and Dan. Each had a unique role in this presentation and conversation, from Edwards's official role documenting the area visually on maps and clarifying the problem and need, to Tom's discussion of his personal connections to the community (both religious and recreational), to Dan's professional ethos of experience with conservation work in another state. Far from being a singular effort of Edwards's recruiting corporate buy-in from the XYZ Company, Edwards recruited Tom and Dan's efforts, as well. Edwards knew that Tom and Dan's ethos and pathos, based on their own experience, would also lend persuasive appeal in recruiting XYZ's involvement in supporting these efforts.

Finally, framing a positive future, the developing new value, was another value that Edwards and the community could agree upon; this perhaps was the most obvious value to align, although this alignment did present some difficulty, based on previous negativity surrounding Grey Cliffs. Initially, the information Edwards received about Grey Cliffs framed the area negatively. While some of its natural beauty remained, the physical degradation of the area, along with the crime statistics documented by the sheriff's office, characterized Grey Cliffs as decidedly lacking in ways that must be improved in order for Grey Cliffs to continue to fulfill the Corps' mission. Conditions at Grey Cliffs were so bad that the public service announcement, designed to apply to activities at all lake-access points, was filmed at Grey Cliffs to serve as the ultimate example of needed improvement. At first, Edwards himself had some difficulty envisioning a positive future at Grey Cliffs given its long-running negative history, and the community, while desiring a positive future for Grey Cliffs, was not sure how that could be accomplished through a relationship with Edwards.

However, despite Edwards's and the community's different positionality regarding Grey Cliffs' history, they could both easily agree that they ultimately *wanted* a positive future for Grey Cliffs. Motivations for improving the area were different: Edwards, while not having as personal of a connection to the area, wanted Grey Cliffs off of his radar as a prominent lake-access point that was causing severe problems. Beyond this, as someone who valued his job and sincerely wanted to promote the Corps' visions and goals, Edwards naturally wanted to pursue an environmentally sustainable area that could be used by compliant members of the public. Also, closing Grey Cliffs, while the easiest, fastest, and perhaps most economical way to address these problems, could also generate negative publicity about the Corps and its relationship to its lands. If closing lake-access points became a precedent, how would others view other problematic lake-access points? What would resolutions to those problems be? Grey Cliffs certainly wasn't unique in the problems manifesting themselves there. It's possible that a precedent could be set at Grey Cliffs that could encourage the closing of other Corps-managed areas that experienced lesser offenses than those experienced at Grey Cliffs, too. Clearly, a positive resolution to this conflict would benefit more than just the Corps and local community members.

Ultimately, the community's history at Grey Cliffs and their initial values of religion, recreation, tradition, and also unity motivated community members to agree on an aligned value of framing a positive future for the area. Behind these values was the community's desire to access Grey Cliffs as it always had, based on these overall community values. This desire motivated the community to negotiate with Edwards to ensure that past problems would be minimized as much as possible. While agreeing to frame a positive future with the new Corps and community plan for improvement, the community needed to agree to the changes that would have to be made to ensure that improvement. Those involved, based on their interviews, indicated the realization that this balance of improvements was tenuous: if community members and outsiders did not follow through with changed behavior, the Corps would need to revisit the problem, and the results next time might not be as positive, since the community had received clear opportunities to do what they could to improve.

The Corps and community did not need to align *all* of their values in order to see improvement and a resolution that "made sense for now" (Weick, 1995). As expected, individual values at the beginning of the conflict emphasized reacting to the current conflict, which had reached a crisis point. As the meetings, collaborative efforts, and one-on-one conversations took place over the course of several months, though, Edwards and the community collaboratively aligned their values to such a degree that the values of sustainability, community buy-in, corporate buy-in, meaningful relationships, compliance/convergence, and framing a positive future contributed to a tentative resolution that appears to be working for now. These aligned values, as reflected in various documents, created space for community participation while preserving Grey Cliffs. In contrast to some conservationist documents that elevate conservation approaches to the point of "exclud[ing] a role for people . . . except to conserve and appreciate nature" (Graham & Lindeman, 2005, p. 442), this collaborative approach included the community's involvement in many significant, visible ways that not only improved the natural environment but also sustained community connections to it. While all of this negotiation took a lot of work, these efforts were not "planned" to align values in a certain way. While Edwards's efforts appeared to be more intentional, based on training he had received as a Corps resource manager, the community's efforts appear to be based on trial and error, a sincere desire to change, and a deep realization of the consequences of contributing to the patterns and behaviors that had brought about Grey Cliffs' downfall. As Edwards and the community worked together to negotiate aligned values through these co-constructed frames, these social actors created the potential to negotiate an aligned ethos as well, since values play such a strong role in ethos development. This aligned ethos and aligned values were necessary for all communicators involved, as they progressed toward resolving this conflict.

An Aligned Ethos

From "the betweens" (LeFevre, 1987; Reynolds, 1993), Edwards and the community co-constructed an aligned ethos based on common values that all could agree upon to varying degrees. This aligned ethos presented a united front that Corps administration and the entire community would need to see to be convinced that improvements actually were occurring in the area. The community represented a diverse group of individuals: while many shared the common values presented in Chapter 3, each person had experienced family time, religious experiences, and traditions differently. In order for Edwards to continue the process of aligning his ethos with the community, he would need to acknowledge the importance of this individuality and reflect it back to the community, addressing social justice concerns in the process (Simmons, 2007; Walton et al., 2019). In response to Edwards's reaching out with the possibility of co-constructing a solution, based on an aligned ethos, the community responded positively, as we can see through the tangible actions that community members took to help improve Grey Cliffs' conditions. Community members truly wanted to help, although some resisters to government authority could quite possibly continue objecting to the developing partnership.

By presenting a co-constructed ethos with the potential for shared alignment, even for a short time, the community and Edwards displayed a united front against many of Grey Cliffs' problems that clearly included Edwards and community members as vital participants. In co-constructing his ethos with the community, Edwards aligned himself with the community audience by emphasizing and framing common interests with the community through character attributes of affinity and establishing himself as a virtuous, sincere person who could be trusted. He also established credibility through experience and expertise. During this process, Edwards focused on parts of his experience that the community could ultimately accept more, which were his observations of the littering and off-roading that the community admitted to participating in. In turn, community members negotiated a reframed ethos so that Edwards could trust them enough to begin following through with some of the proposals made, such as the community's participating in surveilling the area, keeping the area clean, and avoiding off-road vehicle use on Corps property. In the process of accepting the community's "revised ethos," Edwards, interestingly, used the surveillance cameras in a way that would police the use of the area from outsiders and would also monitor the community's activities at the same time. Rather than being offended by the use of the cameras, the community appeared to accept them as just part of the plan to keep the area open. The community accepted the inevitable use of the cameras as a way to acknowledge Edwards's revised, co-constructed ethos, as well: Edwards was willing to give permission for both insiders and outsiders to continue using the area, but the video evidence of that use would be taken as documentation for how well these new efforts were working. The presence of cameras in the area illustrates especially the tentative nature of these efforts: while promises and plans were made on both sides, the evidence had to back up continued success in a way only the Corps would ultimately evaluate, at least from an activities perspective.

This unstable alignment reflects the concept of kairos, "the moment in time when speaking and acting is opportune and when this opportunity has important implications for a concept of agency" (Herndl & Licona, 2007, p. 134). Agential action also hinges on "embodied intuition" (Herder, 2015, p. 353), especially when storytelling within larger narratives, and includes a kairotic, "intuitive understanding of local conditions in order to be able to capitalize on fleeting opportunities and brief openings in dominant narrative strategies" (Herder, 2015, p. 359). In order for Edwards's values and those of the community to align with the potential for social action, the moment had to be opportune. Edwards created that moment at first by suggesting closure, not fully realizing the implications of such a suggestion from the community's side. Because "ethos implies the authority to speak and act with consequences" (Herndl & Licona, 2007, p. 134), Edwards, the one with the most authority, had to assign authority to community members in order for their suggestions to have merit. Likewise, the community had to acknowledge and recognize Edwards's authority so that he would have the ultimate sanctioning power to validate the co-constructed community solutions. This complicated scenario illustrates how agency was enabled and constrained, to varying degrees at different times, throughout this discursive, ethos-building relationship.

THE NEGOTIATED RESOLUTION: ACCESS AND ENVIRONMENTAL PRESERVATION

While both the Corps and community clearly co-constructed agency and negotiated ethos in order to accomplish social action, in some ways, these new narratives of aligned values still support "a dominant social order" (Herndl & Licona, 2007, p. 135) of governmental regulation. The narrative of following rules and regulations never changed, really; the rules and regulations themselves, along with their consequences, never changed. However, at the beginning, the community and other outsiders were not following these rules and regulations at all, so some action had to be taken. In addition, while some governmental and corporate regulation processes attempt to conform the public and employees to align with self-interested goals (Alvesson & Willmott, 2002), those interests seem minimal in this case. Yes, the Corps needed to minimize the negative publicity and potential economically draining lawsuits resulting from community members' not following the rules, and, to some degree, resolving this issue could be a "feather in Edwards's cap" professionally, but, for the most part, the focus of this regulation was on environmental preservation. That Corps goal seemed even more elevated than the goal of providing access to the public, since the original idea for closing the area would limit public access, possibly permanently. This observational case study, then, illustrates opportunities for compromise, as all members participating in this conflict resolution gained something positive for their efforts, although none gained everything they desired from the beginning. Essentially, Edwards and the community created a reciprocal relationship that enlisted all parties in "co-regulating" themselves and outsiders to ensure all were following the rules and preserving this space for all to use far into the future. These citizens gained power and agency through their changing narratives and specific, changed actions, all supported by Edwards. Community members felt "empowered" with these negotiated "tools" (Williams & James, 2009, p. 95) for access. The resulting, negotiated resolution truly did require compromises from both sides, as well as a "wait-and-see," somewhat unstable positionality, as all parties remained poised to see the continued progress of their collaborative, regulatory efforts.

During my interview with Edwards, I asked him if he thought the success of these efforts meant the conflict was over or resolved, at least for now. Edwards replied, "This is not anywhere near over . . . I mean, it's far from over. This is, I think, it's the calm before the storm, personally." While I did not hear any skeptical comments from community members about continuing this positive resolution, Edwards certainly did not seem to expect that an aligned ethos and values could continue on without additional negotiation. As new populations continue to visit Grey Cliffs and as some stories fade into the past, no doubt more conflict will arise in the future, requiring additional alignment in order for everyone to be able to continue using the area. However, even though, as members of the public, "we all reason in the context of a polarized argumentative situation" (Eubanks, 2015, p. 71), arguments such as these about environmental responsibility and sustainability do not need to be presented "as two-sided" (Eubanks, 2015, p. 6) between "privileged technical and vulnerable nonexpert and nonhuman communities" (Olman & DeVasto, 2020, p. 20). Agency negotiation allows space for deconstructing those power imbalances, such as what we see through the results of this case study. Such negotiation moves beyond the deficit, one-way model of communication to a more collaborative, contextual model (Butts & Jones, 2021, p. 10; Druschke & McGreavy, 2016).

The conclusion that follows in the next chapter presents the status of this negotiated resolution 3 years later, as well as a reflection on cycles of agency related to such conflicts and the potential for changing values and sensemaking in the future. The conclusion also discusses limitations and future implications of this case study.

KEY RECOMMENDATIONS FOR TECHNICAL, PROFESSIONAL, AND ORGANIZATIONAL COMMUNICATION AUDIENCES

- Based on interactions with communities and organizations, analyze values from all participants, and highlight the potential for value alignment, where present.
- Acknowledge and encourage discussion of community members' unique positionalities, based on their physical relationship with the environment experiencing risk.
- Assess the role of various types of evidence in conflicts, and demonstrate awareness of the rhetorical context surrounding the use of this evidence.
- Discuss and clarify with audiences why certain participation might not work at a certain time, and recommend what actions could be feasibly taken, as Simmons (2007, p. 9) recommends to researchers involved in environmental decision-making processes.
- Incorporate individual and group stories, including counterstories and antenarratives, to encourage the continued, dynamic process of value alignment.
- Recognize that agreed-upon, new values may develop during conflict resolution and these can be emphasized in new documentation and developing narratives.
- Value various types of expertise and experience when communicating with community members and government officials, rather than just scientific and technical expertise.

KEY RECOMMENDATIONS FOR ENVIRONMENTAL SCIENCE AND PUBLIC POLICY COMMUNICATION AUDIENCES

- Recognize that a focus on compliance could lead to complacency because the compliance becomes the focus, rather than aligned values.
- Encourage convergence, or alignment of common values to the point of desired behavior, to motivate audiences to participate and remain engaged.
- Engage local community members by emphasizing how they can contribute useful information and experience to develop and change policies, based on embodied knowledge. In some cases, environmental communicators may need to "make users aware of their own position/orientation in the space (bodies/boundaries)" (Butts & Jones, 2021, p. 10).
- Rather than focusing only on mainstream technical and scientific knowledge, realize that citizens can possess less formal expertise, at times, that can aid in environmental preservation and conflict resolution.
- Identify other forms of community expertise that can reveal potential for and contribute to new and aligned values; these values can then be reflected in policy development.

8

THE CONTINUED NEGOTIATION PROCESS
Implications for the Future

> "I would say [Grey Cliffs is] evolving, I would say that it . . . has a bright future, I see an area that kids are going to be able to go to and learn from and experience, you know, the great outdoors, and there are several initiatives that we can tie into the community."
>
> —David Edwards, U.S. Army Corps of Engineers resource manager

I waited in my car at the stop sign, anticipating turning left against oncoming traffic onto the main road that led down to Grey Cliffs. Vehicle after vehicle passed me with canoes and kayaks strapped to roof racks or loaded onto trailers. It was summer, and traffic was really picking up in the area. I wondered silently whether the improved conditions and positive social media messaging were responsible for the increased interest in and use of the area. As an outsider, technically, I found myself observing this extra traffic with mixed emotions, since I personally was relatively sheltered from Grey Cliffs' activity now. Locked gates would keep unwanted traffic from our property. However, as an insider—based on my geographic proximity to Grey Cliffs and some history with the area and love of the people who had a stronger history here—I also felt a sense of hope that the community's, the Corps', and the XYZ Company's efforts were working. Aside from some minor communication with Edwards about my research, though, I had not heard any significant word from community members about the work at Grey Cliffs. Interestingly, the formal meetings had ceased, causing me, and probably others involved in the rejuvenation efforts, to feel varying degrees of outsiderness.

The increased traffic in this area also suggests broader possibilities for this research, since more and more outsiders are being drawn to Grey Cliffs, suggesting that eventually some sort of conflict will once again emerge. The successful conflict management analyzed here certainly yields implications for other community members attempting to foster dialogue with organizations such as the Corps, implications that feasibly could impact communities and government representatives far into the future. Likewise, this ethnographic, observational case study demonstrates how successful government representatives' efforts can be in communicating with the public if they connect to those communities' values and ethos. This concluding chapter reflects on the research questions posed in the introduction to this book and discussed throughout the preceding chapters. In reflecting on these questions, I discuss cycles of agency as they relate to Grey Cliffs, the passing of time and its impact on the conflict, the continued potential to negotiate changing values, limitations of this research, the need for continual reflection in conflicts such as these, and future implications.

CYCLES OF AGENCY

In his work on image and narrative, Faber (2002) discusses how Pleasant View, a community cemetery, took control of its image; it had lost its power and agency to control its image, and, instead, external forces were structuring it (p. 162). The cemetery's narrative had moved from a time-honored place that connected with community narratives to a potential financial asset that could be sold to a broader, impersonal corporation that had no connection with the community. However, through Faber's consultant efforts and working with the cemetery's staff, "It [Pleasant View] was able to assert its role within the community and recognize its unique cultural and historical position. By re-aligning its narrative and its image and by regaining control over its external image, the cemetery also regained its power" (Faber, 2002, p. 162). Part of the process of improving Pleasant View's image was connecting it more to the community and its values, including its culture and history. In addition, Faber writes, "By remaining in control of its organizational discourse, Pleasant View was also better able to control its identity by recreating its image of itself" (Faber, 2002, p. 162). This success exemplifies how a community negotiates agency to regain control of an area that possessed significant social capital for it; Pleasant View's reconstructed narrative (assisted through Faber's work) played a crucial role in renegotiating this power, which necessarily involved rhetorical, persuasive efforts.

While Faber's and Grey Cliffs' stories do sound similarly hopeful and optimistic about the future, cycles and narratives of agency can change, and rhetoric does have limits. As Herndl and Licona (2007) write,

> In these institutionalized settings, authoritative practices often reveal a power to stabilize, limit, and control meaning and action. Because it authorizes a rhetor to speak, act, and represent, the authority function often represents and reproduces dominant rhetorical and social relations. As it limits the proliferation of meaning and action, authority can constrain agency. (p. 143)

The Corps as a governmental organization certainly "authorizes a rhetor [i.e., Edwards] to speak, act, and represent" (Herndl & Licona, 2007, p. 143); however, what would happen, for example, if another resource manager took Edwards's place, one who was not willing to negotiate with community members, one who did not feel a need to express sincerity and establish credibility with community members, one who was not willing to take the risk of making the relationship connections that Edwards had? Realistically, not all Corps representatives would have these skills, desires, or patience. Such a change in Corps representation could have a large impact on the ability for community members to negotiate agency with a government institution such as the Corps. Perhaps such an occasion could introduce the opportunity for a new, dominant narrative, which later could "[create] opportunities for new and existing strategy antenarratives to reemerge and struggle for their places in the future" (Cai-Hillon et al., 2015, p. 173). As scholars emphasize, "The process is cyclical" (Cai-Hillon et al., 2015, p. 173). This dynamic reveals the "constant shifting of power" that impacts "self-definition, relationships with oppression, and relationships with others" (Walton et al., 2019, p. 113) occurring among organizational actors and community members.

Likewise, the increased traffic I observed at Grey Cliffs indicates a growing presence of outsiders who could impact the positive, shared narrative and efforts that had improved Grey Cliffs' status so much. Probably not as invested in the area's maintenance as those who live next door to it, these recreational users could have limited access to the established narrative; they might know only what they could read on social media postings about the area, which could present fragments of history and Grey Cliffs' present conditions. As a result, Grey Cliffs could begin experiencing decline again; the new users of the area could quickly outnumber the community members who had invested so much time and effort into preserving the area and promoting the positive narrative that had accomplished so much. In other words, the original community members behind the changes could lose their opportunities to negotiate agency with the Corps; their agency could be diluted through expanded use of Grey Cliffs and the further integration of outsiders into this conflict-resolution process. The various texts and narratives surrounding use of the area would also, therefore, change. The process of negotiating agency requires a give-and-take in co-constructing narratives and antenarratives, including the localized stories within them. An important question remains about all of these efforts: how long will this positive, co-constructed narrative continue? As human nature indicates, a strong tendency exists to always go back to the "way things were" because often those things are more familiar and easy; they require less explanation and effort, and complacency sets in (Sauer, 2010). Quite feasibly, this negotiated resolution could very easily regress back to the Corps' stressing and enforcing the regulations as well as community members' abusing them because the old narrative structures and harmful Grey Cliffs activities could return.

In such an easily imaginable scenario, the Corps and community could experience what Salvo (2006) refers to as "the limit of rhetoric" (p. 229):

> Indeed, often rhetors have to learn to recognize when situations are not rhetorical at all, but are immune to the effects or interventions of rhetoric.... Instead, rhetors must learn when, where, and how they can engage with and intervene within discursive exchanges to make a difference. Such is the limit of rhetoric—that there are indeed rhetorical situations and then there are extrarhetorical or nonrhetorical moments, times when language and symbol manipulation alone will not yield to change or alteration in the space of the community, when one discourse or technology or practice will not change no matter how articulate and well wrought the language brought to bear. (p. 229)

The complex rhetorical discourse and narratives the community and Edwards participated in over time were far from "language and symbol manipulation alone" (Salvo, 2006, p. 229), and yet, a strong potential exists for others to become involved in this negotiated resolution, who could craft this current rhetorical situation into one that no longer offers the potential for negotiation, a co-constructed ethos, or agency. Such is the tenuous nature of the resolution as it currently exists that "makes sense for now" (Weick, 1995). Only time will tell how long this co-constructed resolution will continue working or may require change as users' behavior and activities change over time.

THE CONFLICT, 3 YEARS LATER

As I write this final chapter, almost 3 years after the conflict began, the communication surrounding the dispute itself seems to have died down in some interesting, unexpected ways. When I interviewed Edwards during the winter of 2019 about the future of the negotiated resolution, he had said,

I'd be naive to say that the winter months are the, when we're going to see—no—we're going to see the use increase in the summer, and that's when we're really going to have to start stepping up what we do down there, and I, you know, what I see as a long-term, you know, a volunteer there, they can open and close the gate, or the county, that can, you know, if I ask them to close the gate at night.

Edwards had forecasted the need for strengthened efforts during the summer months due to increased activity. While that activity had certainly increased, Edwards had indicated to me that any compliance concerns had been "relatively quiet," that, while he had expected problems to come back significantly as the area experienced higher use levels, he had really seen only minor issues with compliance, and overall the area had remained clean and crime free. The area was regrowing according to plan, and the conditions had visibly improved so much that more and more people were motivated to visit and enjoy the area. Indeed, Tom's Facebook photos showed cars lined up alongside the road leading down to Grey Cliffs; the area was obviously crowded. In addition, occasionally, the community or a supporting organization had been organizing a cleanup day, and those efforts had been paying off.

While social media communication on YouTube continues about the area's conditions and could be part of the reason for the increased traffic in the area, the community has not continued meeting in organized ways to spur on the improvement efforts. While some community members undoubtedly run into each other at Paul's general store, no signs about community town hall meetings have been posted outside the store, and no formal announcements have been extended to the community as a whole. Paul himself appears at his store less frequently due to health problems. The Grey Cliffs Facebook page that Tom manages has been quiet over the past 2 years. When I interviewed him, Tom expressed surprise that shortly after the community and the Corps agreed upon a tentative resolution, the meetings ceased. He said,

> See, we hadn't had a meeting in so long; I, I really expected someone to say hey, you know, [Frank] or somebody to get ahold of [Edwards] and say hey, we need to try to get a meeting together, I mean, I—and those were the things I thought we would try to do, and I think [Lee] even brought it up, too; I bet, you know, we need to have these [meetings] once every so many months or whatever, and I just, I was really, I'm really shocked, I guess [Frank's] kind of like, well since the Corps kind of stepped forward and going, we don't have to have them [meetings], but, you know, it would have been nice to have a meeting. I didn't know nothing about that [the last cleanup day] until [Frank] posted it on Facebook. And then, you know, then I copied it and pasted it [to my Facebook page]. And I

told [Denise], I said, well, you know, that would have been something we should have had a meeting and posted it, hey we're having a meeting and [put] it up like, you know, the people that can help, come pick up trash or the ones that can, whatever they need to do, you know, everybody could know. And I, you know, I kind of left [Frank] in charge of kind of that, so.

I think we need to have, and I think since the summertime, you know, coming up, we might have more meetings, I think it's 'cause, we really haven't had nowhere during the winter, you know; I know [Paul's] got this back here [a place for people to gather outside], but it's really not big enough, especially if you get as many people we had, you know. And let everybody know, um, we just, there's really nowhere around that we can all just get together and have a meeting like that.

Tom emphasized that for the negotiated resolution to continue, community buy-in and meetings would need to be a part of that process:

> You know, and I think that'll be what will have to happen [have more meetings]. To me, I think we should have to have some more because, you know, you're gonna have a lot of people down there . . . I guess it's just the wrong time of the year, wintertime, so I think this summer a lot of it, a lot more will be coming together.

Paul contributed a similar perspective about needing to hold more meetings in the future:

> I think they still need to keep up the meetings. Yeah, for sure, and let everyone, even though we've got a community, a board, uh, community board, you know, they need, it's their responsibility to keep the rest of the community in the loop. 'Cause there's nothing better than word of mouth. Uh, the meetings are the number one that needs to continue, face-to-face meetings, and they're always welcome here.

Lee confirmed the importance of the meetings, as well, including the need to continue communicating with other community members and the Corps:

> I think we need to continue to communicate to make sure we follow through with what we said we were going to do, and that is clean the area up, upgrade the parking lot, upgrade the camping sites, upgrade the boat ramps, and so on, so I think we need to continue to communicate, you know, the desires and needs of that area.

For all of these community members, the resolution to this conflict was certainly going to be a long-term effort between the community and the Corps. The plan Lee envisioned for fostering a positive relationship between the Corps and community, in which the values of both would be realized, emphasizes the significance of allowing all voices to have access to this discussion. Lee was aware that if the community could not follow through with the plan to regrow the area, the community would lose its access to a much-loved space. The relationship between the Corps and community was the key to this plan's success.

However, by the third summer following the resolution, no formal meetings had been announced. Part of this lack of meetings could relate to the pandemic. Following the interviews I conducted in winter and spring of 2019, one "normal" summer took place before the COVID-19 pandemic hit in 2020. Many people did vacation at lakes and recreational areas during that time because of the social distancing opportunities those outdoor activities (such as swimming, boating, camping, and hiking) provided, but it's possible that Grey Cliffs did not see as many visitors during the summer of 2020 and that, in the summer of 2019, the conflict was so etched in community members' recent memories that the rules and regulations were more at the forefront of people's minds at that point. The summer of 2021, then, could have been the first time for this resolution to be tested.

When I asked Paul about whether he had heard about any future meeting plans, he responded that he had not:

> Actually, it should be time for another meeting. Yeah, no, I haven't heard anything, and I think it's time. But yeah, I was thinking about that when you came in here the other day; is, it's time to have another meeting and find out what's going on here.

Edwards interpreted the lack of meetings and dialogue as a possible indication of lack of support:

> And although, what's interesting is that now that there's no more proposal to close it, I don't have as much of the support that I had going into it, you know, I don't have this much participation, [but] there's a few people that have continued to stick around.

The lack of meetings could simply be a result of contentment, as Tom had mentioned: once the community was convinced of Corps dedication and participation in improving Grey Cliffs in partnership with the community, no one was motivated to continue meeting. The crisis had been averted, and everyone lapsed into a sense of security that the conflict had been resolved, perhaps similar to cycles of attention that Eubanks (2015) and Lester (2010) have discussed regarding environmental issues and the media. This complacency also could relate to an interpreted similar cultural identity the community now saw in relation to the Corps. Hartelius and Browning (2008) write, "When the audience and the rhetor share a cultural identity, they can communicate without making every assumption explicit" (p. 24). Such a perceived similarity would normally take time and continued affirmation through positive interactions over a variety of events and activities, which arguably were taking place at Grey Cliffs. But possibly the community, especially, could have developed such confidence in Edwards that they did not see a need to continue reinforcing community support or continuing close communication with Edwards. Based on Edwards's and community members' narratives, though, this lack of interaction and connection was "interesting" to Edwards and "shock[ing]" to Tom. To somewhat of an outsider like myself, it seemed that the infrastructure was missing to give all of these efforts some consistency. This kind of infrastructure is what Norma had tried to establish by creating a nonprofit 501(c)(3) organization, but the community, not trusting Norma, contentedly accepted a more temporary "status quo" of satisfaction with Corps and community interactions. Most importantly, all were continuing to be able to access Grey Cliffs, although in more restricted ways, and that goal was ultimately what the community desired.

Based on the evidence gained from the meetings, the changed attitudes of Edwards and community members, and the changed physical appearance of Grey Cliffs and plans to continue maintaining that status in a variety of ways for the future, all participants could acknowledge that these interactions and actions had resulted in a new narrative and accompanying texts, albeit evolving ones. This process involved negotiating a more aligned, co-constructed ethos based on values and actions that all could agree on; these values and actions also needed to be somewhat stable and sustainable to carry this resolution into the future. In addition to viewing this negotiation process of narrative and ethos as framing and counterframing, a reciprocal, back-and-forth and give-and-take regarding values, we can view these actions of co-constructing an aligned ethos through the lenses of symbolic capital (Bourdieu, 1986, 1987), agency (Giddens, 1984), and co-constructed value frames. These lenses acknowledge the community's negotiated power with the Corps as community members participated in changed actions. Community members negotiated power rather than just merely complied with Corps regulations.

THE CONTINUED POTENTIAL TO NEGOTIATE CHANGING VALUES

This type of tenuous, unstable resolution suggests that at any time the joint efforts between the Corps and community could to some degree fail. It seems likely that, over time, rules and regulations violations could again be documented via county sheriff reports, and Grey Cliffs visitors could once again begin creeping toward the necessity of more formal, enforced sanctions that could restrict the area's use even more than what everyone had already been experiencing. Such regression might evidence, again, a change in values among Grey Cliffs users; such a change could originate more easily in a diverse group of community members than it would with Corps values: most likely, Corps values would remain the same or could become more stringently associated with the rules and regulations since, ultimately, enforcing those is one of the Corps' primary goals.

The lack of formal communication among community members through meetings could also be seen as a change in community values. Changes in Grey Cliffs' status could cause all parties to revert more to their original values that seemed so polarizing at the beginning of the conflict. However, such changes and failures wouldn't need to be seen in a completely negative light. Potential future failure that some might find inevitable in such a less-than-stable situation can also be framed more positively, a perspective that could frame failure as a way to ensure continued progress in innovative efforts. As Smith et al. (2020) discuss, "Attempts at innovation are often plagued by failure; therefore, innovation is aided by an organization that recognizes and accepts failure in the pursuit of uncertain projects" (p. 5). While the Corps as an organization would frame failure differently in its relationship with the community than it would within its own organization, in the Grey Cliffs case, the community members function as the actual "workers" serving the function of maintaining Grey Cliffs for the Corps. As such, since it would be hard for Edwards to monitor so many community members, failure would likely occur, and Edwards's framing that failure more as opportunities to move forward could create encouragement among community members, rather than framing that failure as shame and discouragement that efforts did not succeed. If the failure were a result of mostly complacency or apathy, though, those changes during attempts to innovate in this situation would be much harder to accept and frame positively.

Referring to their work on framing failure as a resource for continued positive change, Smith et al. (2020)

> demonstrate how these workers communicatively construct the idea of failure as an organizational necessity to develop a schema of action that will contribute to future innovations. When failure is framed in this manner, it effectively neutralizes the contradiction seemingly present in encouraging organizational members to pursue innovative efforts with uncertain outcomes. (p. 3)

Certainly, the Grey Cliffs case presented potential for anxiety as community members could experience a lack of confidence that their improvement efforts would improve the area in ways that met Corps approval. For whatever reason, Edwards's "190-degree flip" that indicated his willingness to consider other options besides closure, according to Tom, demonstrated community support so much that the community was effectively persuaded to move ahead without fear of the "uncertain outcomes" (Smith et al., 2020, p. 3). This confidence could possibly be misplaced since the lack of meetings leaves no opportunities for Edwards's continued affirmation of these efforts' success. However, if the time comes for Edwards to express his affirmation publicly, he could frame any compliance "slips" as opportunities for continued positive change, convergence, and innovation. Smith et al. (2020) continue,

> The role of individuals facilitating innovation is to resource failure as a tool for learning and growth instead of as a signal of unmet goals. When successful resourcing takes place, failure can be seen as a trigger to more attempts at innovation and ongoing learning instead of a call to cease efforts. Resourcing should be understood as constituting, and constitutive of, the organizational communication that supports ongoing innovation efforts. (p. 18)

Both Edwards and the community could be presented with future opportunities to frame Grey Cliffs events as positive rather than only negative: Edwards from his official vantage point as Corps resource manager would have access to different information about Grey Cliffs' status; his framing of crimes in the area, for example, might not be completely positive, but in comparison to past crime statistics, perhaps any new crimes might be interpreted differently. New infractions might not contain the cumulative history, impact, and weight of past crimes, for example. Likewise, community members experiencing Grey Cliffs' environment daily might frame continued littering through the eyes of what others might do to help, such as organizing another cleanup day or installing trash containers, as Tom had suggested. Far from being idealized solutions that ignore the problems at hand, this more positive framing would view any future negative events through the positive frames of what has already been accomplished and future possibilities for success.

This positive framing of failure as opportunity connects to rhetorical uncertainty in risk communication (Sauer, 2003; Walsh & Walker, 2016) that presents uncertainty not as something negative to be avoided but as "a reflexive tool that afford[s] multiple, diverse inventive opportunities" (Walsh & Walker, 2016, p. 71) to develop new knowledge and strategies for managing risk. As Walsh and Walker (2016) point out, the goal when grappling with uncertainty is not for it to be "reduced to zero (or as close thereto as possible)" (p. 72), as technical and scientific perspectives often promote, but to view the range of uncertainties as generative opportunities "for inventing new discourses and new communities around shared risks" (p. 72). This view promotes a future hope for Grey Cliffs and its community, rather than a dread of possible changes that could occur. The multitude of uncertainties surrounding Grey Cliffs and its conditions could generate, just as easily, many positive opportunities for change, bridging the divide between a local citizen group and the "technocratic" Corps of Engineers.

In essence, the community, especially, would be changing its routines and continuing to accept the need to reestablish sensemaking (Weick, 1995) and values that addressed the evolving Grey Cliffs narrative. Referring to Giddens's concept of routinization (Giddens, 1984), Faber (2002) discusses how this concept

> describes the ways routines, habits, and unconscious behaviors align people within and against specific social structures. These routine activities give life its sense of permanence and provide security in the face of change. (p. 142)

Complementing Smith et al.'s (2020) positive, resourcing, framing strategies, Giddens's structuration theory, including the process of developing new routines, establishes the inevitable need to continue interacting with, adjusting to, and navigating social structures, such as the Corps' rules and regulations.

Continuing to Negotiate Values Through Changing Narratives and Stories

For a community that clearly values narrative so much as part of its culture, as evidenced by its consistency of narrating past Corps events as well as stories surrounding Grey Cliffs, the continued prominence of stories and narratives will no doubt continue to play a role in framing the communication of values and sensemaking of the larger Grey Cliffs narrative, including developing texts that continue reflecting these values. Smith et al. (2020) indicate that "Stories are helpful for providing an extended relatable example and building rapport" (p. 6). Continuing these stories will be essential to framing the Grey Cliffs' future narrative positively and maintaining community relationships through this common focus and efforts. In their work, Smith et al. (2020) found that

> the findings not only highlighted the significance of being able to tell stories that sell innovation internally and externally, but also the skill to tell stories that construct failure in a way that minimizes its negative connotations while also encouraging others to pursue novel ideas. (p. 19)

Likewise, in his work focusing on organizational and community change, Faber (2002) stated,

> I witnessed and experienced the ways people use stories and myths within unstable, changing contexts, and I gained many insights that helped me see the connections and the order communication can build when our social spaces, our work spaces, and even our religious spaces become threatened. (p. 141)

Edwards's and community members' narratives indicate the crucial role these stories of Grey Cliffs' past contributed to the potential for unifying action among all of those involved; the narratives and antenarratives communicated evidence of past loyalties and values, as well as the eventual desire to change in the face of threats expressing the potential for imminent loss of access to this geographic space. The stories narrating evidence of positive change also created cohesion within this community as it partnered to oppose the original Grey Cliffs narrative that did indeed threaten the essence of this community and its values.

Continuing to Negotiate Values Through Ethos Building

Notably, rhetoric will continue to play a role in the interactions with this changing landscape; actors will continue to co-construct agency, ethos, and meaning. As activities surrounding Grey Cliffs evolve and change, Edwards and the community will need to continue their collaborative communication and physical efforts; current efforts may fail or become less effective over time, and these social actors will need to continue "to pursue novel ideas" (Smith et al., 2020, p. 19). If the Corps and community can continue this positive framing strategy while also acknowledging that sensemaking and values will undoubtedly change over time, all parties will be able to view these innovations within the larger cultural framework that surrounds them and will then be able to accept needed change with this big picture in mind. Narratives and stories will continue to play a large part in co-constructing ethos. As Smith et al. (2020) emphasize in elaborating on Fairhurst's work (2005),

> Managers are not transmitting information; they are "managers of meaning" (p. 166) who co-construct a shared reality for others. By utilizing the language tools of metaphor, jargon/catchphrases, contrast, spin, and stories, individuals can actively craft the reality of situations. (p. 6)

All of this language work is necessarily rhetorical and requires not only the work of managers such as Edwards but also community efforts; through stories and daily communicative efforts, these community members negotiate meaning with insiders, outsiders, Corps representatives, and countless others in between. Considering ethos as a "spatial metaphor," as one of the "language tools" that Smith et al. (2020) also elevate as part of the important work of "actively craft[ing] the reality of situations" (p. 6), seems especially appropriate when discussing the negotiated reality of Grey Cliffs as a geographic space: the physical reality of accessing this space is actively being negotiated every day through ethos construction revealed in the constant creation of narratives, stories, and texts that reflect the impact of the past as well as present and future possibilities for continuing to renegotiate community interaction with this area.

Scholars continue to highlight future possibilities for rhetoric's role in organizational research such as this:

> We suggest that management research conceptualize rhetoric as a theoretical lens focused on organizational interactions, as well as a practical mode of intervening in those interactions; the substance that maintains organizational order and institutional logic, as well as the means of challenging that order; a producer and facilitator of individual and organizational identity; the manager's major strategy for persuading followers to enact management philosophies; and a framework for understanding the role of narrative and rational organizational discourses. (Hartelius & Browning, 2008, p. 33)

While Edwards's role as manager in persuading this community to act cannot be minimized, the success of these efforts could not have been possible without Edwards's openness to and negotiation with the community members in his managerial role. In turn, these community members clearly expressed a desire to compromise and align their values, as well as their individual and collaborative ethos, in order to achieve a negotiated resolution with the Corps. While all parties seem very happy with the negotiated resolution, the absence of a more formalized plan among community members to continue negotiating when future problems arise remains conspicuous.

Edwards's collaborative, rhetorical efforts, represented by the revised narratives and other texts he used to communicate with the community, worked to encourage the community to accept Corps values of sustainability, trust, ethical behavior, and relationship, eventually. While some might object to the "panoptic" (Foucault, 1977) view of the Corps in maintaining compliance through hidden video surveillance, for example, which Edwards had acknowledged, the community seemed to accept this need to negotiate with the Corps in necessary ways to keep their access to an area they loved. From my perspective as an insider (witness to these events), outsider (not being as impacted by the cultural history of the area since I am not from the area originally), and researcher (drawing on qualitative, ethnographic research techniques and interviews with community members), I could see that Edwards ultimately facilitated a healthy communication dynamic that seemed to work well for this community as a whole. I want to emphasize here that my analysis of Edwards's work should not "reif[y] the individual" (Herndl & Licona, 2007, p. 139) by attempting to uphold him as a leader who facilitated this success alone. In contrast, recognizing what did not work at first, Edwards negotiated his ethos with the community as active participants, related more personally to the community through efforts to build relationship, and shared the enforcement of the rules and regulations in a way the community positively accepted and reciprocated overall as he encouraged the community to work within Corps values while also addressing community values. The community members played a strong role in the success of these rhetorical efforts.

Continuing to Negotiate Values Through Environmental and Social Justice Frames

This community involvement, through evolving narratives about Grey Cliffs' solutions, specific behavior changes in a variety of areas such as social media and cleanup efforts, and willingness to compromise and align values and ethos with the Corps, reveals positive, current results of what can happen when organizational representatives make sincere efforts to involve the public in addressing environmental concerns such as these. Certainly, such involvement is not always the case; often, "public participation has fallen short of the ideal model" (Tillery, 2019, p. 10; see also Simmons, 2007), but this work presents some hope for this type of collaboration. While Corps and community values will continue to change and most likely never will fully align, scholars indicate that groups with diverse values can work together on "common environmental causes" (Tillery, 2019, p. 14) and that government officials and publics can identify with each other, build bridges, and consider a range of environmental discourses that allow opportunities for making connections (Prelli & Winters, 2009; Tillery, 2019). Prelli and Winters (2009) suggest that continued rhetorical efforts, such as even "explor[ing] ways to generate new terms in common" (p. 240) make connections through language, such as Edwards did to some degree by introducing the Grey Cliffs community to Corps value terms. This type of collaboration contributes to "rhetorical coalition building," particularly when these terms are part of "story elements underlying environmental discourses" (Prelli & Winters, 2009, p. 240). Because "institutions are dynamic, uncontained structures and as such offer the space/possibility for such change" (Simmons, 2007, p. 10), these "story elements" (Prelli & Winters, 2009, p. 240) from an organizational perspective are key to revealing values and their potential for negotiation and alignment with engaged publics' narratives and stories regarding environmental sustainability. Likewise, community stories can also compel organizational representatives to work with willing community participants, as we see in this case study.

LIMITATIONS OF THIS STUDY

This ethnographic, qualitative, observational case study presents several limitations. One is that this study reflects a social constructionist model of environmental risk communication that privileges the co-production and negotiation of knowledge when solving Grey Cliffs' environmental problems, as well as the social problems experienced by this community as a result of conflicts with Grey Cliffs' conditions and the Corps. However, as Olman and DeVasto (2020) point out, the "ideal" social constructionist model does contain some significant problems when discussing environmental risk communication, such as tending to move more toward a one-way communication model at times, due to the "technocratic framing of the risk situation" (Olman & DeVasto, 2020, p. 17) that emphasizes knowledge differences between experts and nonexperts and ultimately tends to favor the experts with their scientific expertise. Another weakness of the social constructionist communication model is focusing on human agents and agency rather than including more nonhuman concerns, such as the Grey Cliffs environment. Although I discuss the Grey Cliffs environment here, the case study's focus, due to its sociological theoretical foundations and framework, disproportionately highlights human agents and their concerns.

This type of ethnographic research also focuses on in-depth data gathering for a very specific situation. This case study highlights a relatively small community, as well as only one organizational representative. Generalizing the results of this study to more broad populations is therefore not possible; generating generalized strategies was not the purpose of this study. However, when considering these results, researchers and others should realize that individuals and communities are very different; even if the results were generalizable, researchers and practitioners would need to ensure that applications fit the needs and goals of other, localized communities, including their individual and collective cultures. Such an orientation indicates awareness that there are no one-size-fits-all recommendations for this type of collaboration and accepts the individualized, unique locales and individuals involved in such conflicts. This case study demonstrates the importance of historical context when resolving conflict and that other publics engage in conflict with various historical contexts in mind, as well. Likewise, community members' cultural backgrounds might influence communication in general about conflicts, based on tradition. Those seeking to apply this research to other contexts need to keep in mind and address the broader communication contexts surrounding the conflict and adapt strategies accordingly.

Similarly, this case study presents a limited snapshot of this conflict among a few participants during a defined timeframe. The conflict itself took place during a period of several months, although the efforts related to it continue until this day. This research highlights the conflict itself as well as the events and communication surrounding it rather than the continued efforts to prevent additional conflict; discussing continued efforts outside of my defined research frame is beyond the scope of this current work but could provide additional, future research opportunities. In addition, the number of community members actively participating in the town hall meetings was relatively small, and, therefore, the pool of participants to choose from to interview was also small. Participants also needed to consent for me to interview them. While most participants I asked to interview did agree to participate and consented with no problems, I had some difficulty getting in touch with some other potential interviewees. I did not have contact information for Frank; his daughter and her husband, whom I thought would have some useful information, did not respond to my request to interview. While the number of interview participants seemed somewhat limited, the ones I did interview provided detailed information about their connections to the area as well as the conflict in general. Their input was incredibly useful to my research into this community, its culture, and its perspectives on Grey Cliffs, and I am thankful that, while limited, their stories have contributed to developing a record of this conflict and its progress.

My research also focuses on just one interview with participants. Quite possibly, these interviewees could have changed their perspectives on the conflict over time, as they observed the changes at Grey Cliffs. Uncovering aspects of identity and overall cultural perspectives can be very difficult to accomplish during one interview, and these short interactions also are limitations of this research. Because of the short duration of the conflict and the few town hall meetings that occurred, my observations are also limited to the number of town hall and committee meetings held; I also did not attend cleanup days or the meetings with the XYZ Company, so my note taking and observations focus on my limited participation as one researcher and my choice of theoretical lenses through which I viewed and analyzed this conflict and interactions with participants, based on observations and reflective interviews.

While my outsider status might also have been a limitation, it also could have been a positive characteristic, since I was a bit more emotionally removed from these circumstances. My sense was that the community members were eager to tell me the full background of this situation, "the full story"; community members had no preconceived notions about me, since they did not know me before the conflict. This distance could have contributed to my greater objectivity about this conflict and tentative resolution. However, I also have been influenced by my own personal biases that could inhibit objectivity, such as interest in governmental communication and a desire not to alienate research participants, such as Edwards, by overly criticizing his actions and communicative responses within this community. My status as a traditionally educated outsider may also make me more tolerant of accepting routine, dominant discourses of power and be less likely myself to accept, understand, and even hear the voices from nondominant groups, which I sincerely tried to represent to the best of my ability. Characterizing this group as nondominant also presents questions and implications, such as why working-class, perhaps less formally educated community members would be considered nondominant to begin with. I have not fully explored this issue in this research. Likewise, while I acknowledge the discourses of power that Edwards accessed as a governmental, organizational representative, analyzing and critiquing why these particular discourses are granted privilege from the outset is also somewhat beyond the scope of the research I present here.

THE NEED FOR CONTINUED REFLECTION

Whether the community realized it or not, self-reflection was an important part of the agency-negotiation process, and the interviews—the community's self-reflective narratives—emphasized that process during the ethnographic case study research process. As Faber writes, "Self-reflection is structural in that it must be learned, practiced, and socialized. At the same time, self-reflection is an act of agency as it enables the person to see how actions, beliefs, and motives are influenced by structure" (Faber, 2002, p. 122). These community members realized, through the process of meeting and organizing, what Edwards's goals were for Grey Cliffs, the Corps rules and regulations, as well as the consequences of noncompliance. Once they understood these governmental, organizational structures, community members could then collaborate on solutions that worked within these structures, and reflection was a part of this process. The community first had to look beyond their initial reactions to Edwards's proposal to close the area; these reactions were far from reflective and did not indicate any willingness to negotiate. At first, emotion clouded the community's understanding about the power and agency it actually could negotiate with Edwards to arrive at a solution that all could accept. Once Edwards indicated his willingness to compromise, the community was then able to make sense of what was happening, and through the collaborative meeting discussions, they took ownership of the problems in ways that moved the conflict resolution forward. The town hall meetings, then, served as reflection opportunities: community members could discuss what had happened, strategize about future plans, and coordinate efforts. Additional meetings reflected on the success of these efforts and provided opportunities to discuss needed improvements for the future. The narrative interviews I conducted also provided opportunities for reflection. It was during those conversations that community members reflected on Edwards's sudden shift from presenting the idea to close Grey Cliffs to a willingness to collaborate on alternative solutions. Community members also reflected on the dissonance Norma introduced into the situation, as well as the progress on current efforts.

This community needs ongoing reflection opportunities, though, in order for these structural changes, this negotiated agency and ethos, to continue. As Faber (2002) writes,

> If currently powerful agents are unable to maintain the conditions by which their power is naturalized, their status will erode and fade as new agents replace the old order with their own social infrastructures. This is where Giddens inserts agency into power's structures, arguing that agents must choose to replicate social relations in order to build and maintain power. If no one replicates domestic relations, social patterns, or local customs, there is no power. (p. 120)

Without continued reflection, this community runs the risk of the old structures of environmental degradation and criminal activity creeping back in; very quickly, "new agents" (or old agents, in this case) could begin destroying the new, productive power structures community members had negotiated and replace them with the old structures, or, more accurately, with a lack of functioning structure. Community complacency could suggest that community members are "currently powerful agents" (Faber, 2002, p. 140), when in reality, the community as a whole could quickly lose their negotiating capacity, if the once dominant, destructive forces take control once again.

While the community might have difficulty creating opportunities for continued reflections through imposing an infrastructure of future meetings, for example, Edwards, in his role as Corps resource manager, already has that reflection infrastructure built in through the process of completing after-action reviews, which encourage him to reflect on events such as these involving public interactions. Required by the Corps, these reviews require a discussion of what happened, how the Corps representative responded, the perceived reaction of audience members and participants, and any plans to make changes for future interactions, based on what transpired. Edwards seemed somewhat conditioned by the requirement of such reflective thought for his job and mentioned that

> one of the four things I try to always do is, you know, the self-reflection, how did I conduct myself in that meeting, is the message that I was, was I effective in the message that I was trying to give.

While Edwards had presented the idea of closure as an option to indicate the seriousness of the need to respond to Grey Cliffs' problems, he also said that he didn't necessarily want to close the area; that was just one of the options available, perhaps one of the easiest, most cost-effective, and most efficient ones. In reflecting upon that initial proposal to close the area, though, he acknowledged that he felt he had failed, largely because of the strong community response against the idea that Edwards felt may have damaged his relationship with the community members even before a relationship could begin. Not knowing in advance who would be at the meeting or what the reaction might be to closure, Edwards took a risk, then changed his approach to the problem in real time. While Edwards didn't clarify how this experience might influence his response to future, possibly similar conflicts, the fact that he is thoughtfully reflecting on the situation could lead to increased opportunities for communicative flexibility in the future, whether interacting with the Grey Cliffs community or communities surrounding other lake-access points that he manages. Requiring such reflection appears to be an effective Corps strategy, although its long-term effectiveness would depend upon the willingness of organizational communicators to reflect deeply enough to identify future improvements, as well as actually enact those improvements in future rhetorical situations.

FUTURE IMPLICATIONS

The Grey Cliffs conflict emphasizes the essential role of collaboration between organizational communicators and their publics. Far from being experts that hand down regulations and knowledge from above as sovereign communicators in a vacuum, these communicators truly need public participation in order to identify and negotiate effective resolutions. Salvo (2006) writes,

> Ultimately, the role of the expert is shifting. . . . Experts are being asked by citizens to consult with communities to create meaningful, usable, intelligent solutions that are built with attention to local practices, constraints and economic boundaries, from building *for* users to building *with* users. (p. 224)

In his role as "expert" in this case, Edwards, although he had some ideas for ways to resolve Grey Cliffs' problems, needed community members to join with him as experts in their own right; they knew best how to motivate and organize themselves in order to identify solutions that worked best for them and the area at this moment, through their localized experiences. Their cultural-historical knowledge, experiences, stories, unique values, and relationships allowed them to join with Edwards in their jointly constructed solutions that no one could have achieved individually or through an "us-versus-them" mentality. The community needed Edwards, as well; even though they rejected Edwards at first, he possessed knowledge as an organizational, governmental representative that was crucial to the community's efforts; ultimately, these efforts required Corps approval, and Edwards was the mediator who could ensure that happened. Significantly, these social actors co-constructed persuasion through co-constructing ethos in a dynamic, evolving, recursive process that jointly created these workable solutions.

Far from these efforts' being "anywhere near over," in Edwards's words, these interactions have created a foundation and history for future interactions between the Corps and community, a foundation that current participants in this conflict as well as newcomers can build upon, since they are based in a developing, common history. The Grey Cliffs case study facilitates the documentation of such important efforts. Future success in managing this conflict will depend on continuing to understand and negotiate common values, as well as a common ethos.

Referring to types of potentially dynamic communicative situations such as these, Cyphert (2010) states, "A sustainable global society will require robust rhetorical systems with which to make increasingly complex collective decisions that simultaneously involve economic, environmental, social, political, and moral issues" (p. 360). Such rhetorical systems no doubt are enhanced by careful, diplomatic, organizational communicators who address community and stakeholder concerns as well as environmental ones. Astute communicators will realize and address the fact that, in addition to environmental sustainability issues impacting a community, for example, other sustainability needs exist as well, such as the need to sustain communication and relationship channels between a government organization and community members, as well as a community's being able to sustain itself through a continued connection to a beloved geographic space in need of beautification and preservation. Various sustainability needs may require attention within the same physical, relational, and rhetorical context, addressing multiple communities, individual identities, and relationships. Additional research can continue to focus on the multiple voices (Castelló et al., 2013) involved in conflict surrounding sustainability issues and discover more of the "available means of persuasion" (Aristotle, ca. 367–347, 335–323 B.C.E./2019) and identification (Burke, 1969) needed for each community and rhetorical situation.

REFERENCES

Agboka, G. (2013). Participatory localization: A social justice approach to navigating unenfranchised/disenfranchised cultural sites. *Technical Communication Quarterly*, 22(1), 28–49. https://doi.org/10.1080/10572252.2013.730966

Allen, M. W., Walker, K. L., & Brady, R. (2012). Sustainability discourse within a supply chain relationship: Mapping convergence and divergence. *Journal of Business Communication*, 49(3), 210–236. https://doi.org/10.1177/0021943612446732

Alvesson, M. (2004). *Knowledge work and knowledge-intensive firms*. Oxford University Press.

Alvesson, M., Ashcraft, K., & Thomas, R. (2008). Identity matters: Reflections on the construction of identity scholarship in organization studies. *Organization*, 15(1), 5–28. https://doi.org/10.1177/1350508407084426

Alvesson, M., & Willmott, H. (2002). Identity regulation as organizational control: Producing the appropriate individual. *Journal of Management Studies*, 39(5), 619–644. https://doi.org/10.1111/1467-6486.00305

Anand, N., Gardner, H., & Morris, T. (2007). Knowledge-based innovation: Emergence and embedding of new practice areas in management consulting firms. *The Academy of Management Journal*, 50(2), 406–428. https://doi.org/10.5465/amj.2007.24634457

Anson, C., & Forsberg, L. L. (2002). Moving beyond the academic community: Transitional stages in professional writing. In T. Peeples (Ed.), *Professional writing and rhetoric: Readings from the field* (pp. 388–409). Longman.

Aristotle. (1990). Rhetoric (W. R. Roberts, Trans.; F. Solmsen, Ed.). In P. Bizzell & B. Herzberg (Eds.), *The rhetorical tradition: Readings from classical times to the present* (pp. 151–194). St. Martin's Press. (Original work published ca. 367–347, 335–323 B.C.E.)

Aristotle. (2007). *On rhetoric: A theory of civic discourse* (G. A. Kennedy, Trans.; 2nd ed.). Oxford University Press. (Original work published ca. 367–347, 335–323 B.C.E.)

Aristotle. (2012). *Nicomachean ethics* (R. C. Bartlett & S. D. Collins, Trans.). University of Chicago Press. (Original work published ca. 350 B.C.E.)

Aristotle. (2019). *Art of rhetoric* (R. C. Bartlett, Trans.). University of Chicago Press. (Original work published ca. 367–347, 335–323 B.C.E.)

Arnett, R. C. (2002). Paulo Freire's revolutionary pedagogy: From a story-centered to a narrative-centered communication ethic. *Qualitative Inquiry*, 8(4), 489–510. https://doi.org/10.1177/10778004008004006

Bakhtin, M. M. (1983). *The dialogic imagination: Four essays*. (M. Holquist, Ed. & Trans., & C. Emerson, Trans.). University of Texas Press.

Bakhtin, M. M. (1984). *Problems of Dostoevsky's poetics*. (C. Emerson, Ed. & Trans.). University of Minnesota Press.

Baniya, S., & Chen, C. (2021). Experiencing a global pandemic: The power of public storytelling as antenarrative in crisis communication. *Technical Communication*, 68(4), 74–87.

Baumlin, J. S., & Meyer, C. A. (2018). Positioning ethos in/for the twenty-first century: An introduction to *Histories of Ethos*. *Humanities* 7(3), 78, 1–26. https://doi.org/10.3390/h7030078

Bergan, D. E. (n.d.). *Encyclopedia Britannica*. Retrieved January 12, 2023, from https://www.britannica.com/topic/grassroots

Blair, C. (2001). Reflections on criticism and bodies: Parables from public places. *Western Journal of Communication*, 65(3), 271–294. https://doi.org/10.1080/10570310109374706

Boje, D. M. (2001). *Narrative methods for organizational and communication research*. Sage.
Boje, D. M. (2008). *Storytelling organizations*. Sage.
Boje, D. M. (Ed.). (2015). *Storytelling and the future of organizations: An antenarrative handbook*. Taylor and Francis.
Bondi, M., & Yu, D. (2019). Textual voices in corporate reporting: A cross-cultural analysis of Chinese, Italian, and American CSR reports. *International Journal of Business Communication*, 56(2), 173–197. https://doi.org/10.1177/2329488418784690
Bourdieu, P. (1986). The forms of capital. (R. Nice, Trans.). In J. Richardson (Ed.), *Handbook of theory and research in the sociology of education* (pp. 241–258). Greenwood Press.
Bourdieu, P. (1987). *Distinction: A social critique of the judgement of taste*. (R. Nice, Trans.). Harvard University Press.
Bourdieu, P. (1990). *The logic of practice*. Stanford University Press.
Bourdieu, P. (2007). *Outline of a theory of practice*. (R. Nice, Trans.). Cambridge University Press.
Boussebaa, M., & Brown, A. D. (2017). Englishization, identity regulation and imperialism. *Organization Studies*, 38(1), 7–29. https://doi.org/10.1177/0170840616655494
Boyd, J., & Waymer, D. (2011). Organizational rhetoric: A subject of interest(s). *Management Communication Quarterly*, 25(3), 474–493. https://doi.org/10.1177/0893318911409865
Braet, A. C. (1992). Ethos, pathos, and logos in Aristotle's *Rhetoric*: A re-examination. *Argumentation*, 6(3), 307–320. https://doi.org/10.1007/BF00154696
Britt, E. C. (2006). The rhetorical work of institutions. In J. B. Scott, B. Longo, & K. V. Wills (Eds.), *Critical power tools: Technical communication and cultural studies* (pp. 133–150). State University of New York Press.
Burke, K. (1968). *Language as symbolic action*. University of California Press.
Burke, K. (1969). *A rhetoric of motives*. University of California Press.
Butts, S., & Jones, M. (2021). Deep mapping for environmental communication design. *Communication Design Quarterly*, 9(1), 4–19. https://doi.org/10.1145/3437000.3437001
Cai-Hillon, Y., Boje, D. M., & Dir, C. (2015). Strategy as antenarrative complexity. In D. M. Boje (Ed.), *Storytelling and the future of organizations: An antenarrative handbook* (pp. 163–175). Taylor & Francis.
Campbell, G. (1990). The philosophy of rhetoric. In P. Bizzell & B. Herzberg (Eds.), *The rhetorical tradition: Readings from classical times to the present* (pp. 749–795). St. Martin's Press. (Original work published 1776)
Campbell, K. K., Huxman, S. S., & Burkholder, T. A. (2015). *The rhetorical act: Thinking, speaking, and writing critically* (5th ed.). Cengage.
Carlson, E. B., & Caretta, M. A. (2021). Legitimizing situated knowledge in rural communities through storytelling around gas pipelines and environmental risk. *Technical Communication*, 68(4), 40–55.
Castelló, I., Morsing, M., & Schultz, F. (2013). Communicative dynamics and the polyphony of corporate social responsibility in the network society. *Journal of Business Ethics*, 118, 683–694. https://doi.org/10.1007/s10551-013-1954-1
Chávez, K. R. (2013). *Queer migration politics: Activist rhetoric and coalitional possibilities*. University of Illinois Press.
Cheng, M. S. (2012). Colin Powell's speech to the UN: A discourse analytic study of reconstituted "ethos." *Rhetoric Society Quarterly*, 42(5), 424–449. https://doi.org/10.1080/02773945.2012.704121
Christensen, L. T., Morsing, M., & Thyssen, O. (2011). The polyphony of corporate social responsibility: Deconstructing accountability and transparency in the context of identity and hypocrisy. In G. Cheney, S. May, & D. Munshi (Eds.), *Handbook of communication ethics* (pp. 457–474). Routledge.
Convention of States. (n.d.). *What's a convention of states anyway?* Retrieved January 12, 2023, from https://conventionofstates.com

Corbin, J., & Strauss, A. (2015). *Basics of qualitative research: Techniques and procedures for developing grounded theory* (4th ed.). Sage.

Cornelissen, J., & Werner, M. (2014). Putting framing in perspective: A review of framing and frame analysis across the management and organizational literature. *The Academy of Management Annals, 8*(1), 181–235. https://doi.org/10.5465/19416520.2014.875669

Cowan, J. (2017). Topology and psychoanalysis: Rhe-torically restructuring the subject. In L. Walsh & C. Boyle (Eds.), *Topologies as techniques for a post-critical rhetoric* (pp. 151–174). Palgrave Macmillan.

Cyphert, D. (2010). The rhetorical analysis of business speech: Unresolved questions. *Journal of Business Communication, 47*(3), 346–368. https://doi.org/10.1177/0021943610370577

DeKay, S. (2011). When doing what's right becomes messy. *Business Communication Quarterly, 74*(4), 412–414. https://doi.org/10.1177/1080569911423965

Delgado, R. (1989). Storytelling for oppositionists and others: A plea for narrative. *Michigan Law Review, 87*(8), 2411–2441. https://doi.org/10.2307/1289308

Denzin, N. K., & Lincoln, Y. S. (2018). *The Sage handbook of qualitative research*. Sage.

Druschke, C. G. (2018). Agonistic methodology: A rhetorical case study in agricultural stewardship. In C. Rai & C. G. Druschke (Eds.), *Field rhetoric: Ethnography, ecology, and engagement in the places of persuasion* (pp. 22–42). The University of Alabama Press.

Druschke, C. G., & McGreavy, B. (2016). Why rhetoric matters for ecology. *Frontiers in Ecology and the Environment, 14*(1), 46–52. https://doi.org/10.1002/16-0113.1

Dunn, T. R. (2019). *Talking white trash: Mediated representations and lived experiences of white working-class people*. Routledge.

Durá, L. (2018). Expanding inventional and solution spaces: How asset-based inquiry can support advocacy in technical communication. In G. Y. Agboka & N. Matveeva (Eds.), *Citizenship and advocacy in technical communication: Scholarly and pedagogical perspectives* (pp. 23–39). Taylor & Francis.

Edwards, J. (2018). Race and the workplace: Toward a critically conscious pedagogy. In A. M. Haas & M. F. Eble (Eds.), *Key theoretical frameworks: Teaching technical communication in the twenty-first century* (pp. 268–286). Utah State University Press.

Eichberger, R. (2019). Maps, silence, and Standing Rock: Seeking a visuality for the age of environmental crisis. *Communication Design Quarterly, 7*(1), 9–21. https://doi.org/10.1145/3331558.3331560

Eubanks, P. E. (2015). *The troubled rhetoric and communication of climate change: The argumentative situation*. Routledge.

Faber, B. D. (2002). *Community action and organizational change: Image, narrative, identity*. Southern Illinois University Press.

Faber, B. D. (2007). Discourse and regulation: Critical text analysis and workplace studies. In M. Zachry & C. Thralls (Eds.), *Communicative practices in workplaces and the professions: Cultural perspectives on the regulation of discourse and organizations* (pp. 203–218). Baywood Publishing Company.

Fairhurst, G. T. (2005). Reframing the art of framing: Problems and prospects for leadership. *Leadership, 1*(2), 165–185. https://doi.org/10.1177/1742715005051857

Fernheimer, J. W. (2016). Confronting Kenneth Burke's antisemitism. *Journal of Communication & Religion, 39*(2), 36–53. https://uknowledge.uky.edu/wrd_facpub/3

Foucault, M. (1977). *Discipline and punish*. Vintage.

Foucault, M. (1980). *Power/knowledge: Selected interviews and other writings 1972–1977* (C. Cordon, L. Marshall, J. Mepham, & K. Soper, Trans.). Pantheon.

Foucault, M. (1983). The subject and power. In H. L. Dreyfus & P. Rabinow (Eds.), *Michel Foucault: Beyond structuralism and hermeneutics* (pp. 208–226). University of Chicago Press.

Foucault, M. (1995). *Discipline and punish: The birth of the prison* (A. Sheridan, Trans.; 2nd ed.). Random House.

Frost, E. A. (2018). Apparent feminism and risk communication: Hazard, outrage, environment, and embodiment. In A. M. Haas & M. F. Eble (Eds.), *Key theoretical frameworks: Teaching technical communication in the twenty-first century* (pp. 23–45). Utah State University Press.

Gaard, G. (2018). *Critical ecofeminism*. Lexington Books.

Gaitens, J. (2000). Lessons from the field: Socialization issues in writing and editing internships. *Business Communication Quarterly*, *63*(1), 64–76. https://doi.org/10.1177/108056990006300108

Gee, J. P. (2004). *Situated language and learning: A critique of traditional schooling*. Routledge.

Gee, J. P. (2005). Semiotic social spaces and affinity spaces. In D. Barton & K. Tusting (Eds.), *Beyond communities of practice: Language, power and social context* (pp. 214–232). Cambridge University Press.

Gee, J. P. (2014). *An introduction to discourse analysis: Theory and method* (4th ed.). Routledge.

George, B., & Manzo, H. (2022). The Ohio river: Re-imagining water risk through embodied deliberation. In S. Stinson & M. Le Rouge (Eds.), *Embodied environmental risk in technical communication* (pp. 99–118). Taylor & Francis.

Gephart, R. P. (2007). Hearing discourse. In M. Zachry & C. Thralls (Eds.), *Communicative practices in workplaces and the professions: Cultural perspectives on the regulation of discourse and organizations* (pp. 239–263). Baywood Publishing Company.

Gergen, K. J. (2007). Writing and relationship in academic culture. In M. Zachry & C. Thralls (Eds.), *Communicative practices in workplaces and the professions: Cultural perspectives on the regulation of discourse and organizations* (pp. 113–129). Baywood Publishing Company.

Gherardi, S., & Nicolini, D. (2002). Learning in a constellation of interconnected practices: Canon or dissonance? *Journal of Management Studies*, *39*(4), 419–436. https://doi.org/10.1111/1467-6486.t01-1-00298

Giddens, A. (1984). *The constitution of society: Outline of the theory of structuration*. University of California Press.

Giddens, A. (1991a). *The consequences of modernity*. Stanford University Press.

Giddens, A. (1991b). *Modernity and self-identity: Self and society in the late modern age*. Polity.

Glaser, B. G., & Strauss, A. L. (1967). *The discovery of grounded theory: Strategies for qualitative research*. Aldine.

Goffman, E. (1974). *Frame analysis: An essay on the organization of experience*. Harvard University Press.

Goggin, P. N. (Ed.). (2009). *Rhetorics, literacies, and narratives of sustainability*. Routledge.

Golden, J. L., Berquist, G. F., Coleman, W. E., & Sproule, J. M. (2011). *The rhetoric of western thought*. Kendall-Hunt.

Grabill, J. T. (2006). The study of writing in the social factory: Methodology and rhetorical agency. In J. B. Scott, B. Longo, & K. V. Wills (Eds.), *Critical power tools: Technical communication and cultural studies* (pp. 151–170). State University of New York Press.

Grabill, J. T., Leon, K., & Pigg, S. (2018). Fieldwork and the identification and assembling of agencies. In C. Rai & C. G. Druschke (Eds.), *Field rhetoric: Ethnography, ecology, and engagement in the places of persuasion* (pp. 193–212). The University of Alabama Press.

Grabill, J. T., & Simmons, W. M. (1998). Toward a critical rhetoric of risk communication: Producing citizens and the role of technical communicators. *Technical Communication Quarterly*, *7*(4), 415–441. https://doi.org/10.1080/10572259809364640

Graham, M. B., & Lindeman, N. (2005). The rhetoric and politics of science in the case of the Missouri river system. *Journal of Business and Technical Communication*, *19*(4), 422–448. https://doi.org/10.1177/1050651905278311

Grant, K. (2015). The creative spirit of the leader's soul: Using antenarratives to explain metanoia experiences. In D. M. Boje (Ed.), *Storytelling and the future of organizations: An antenarrative handbook* (pp. 101–116). Taylor & Francis.

Griffin, F. (2009). Merck's open letters and the teaching of ethos. *Business Communication Quarterly, 72*(1), 61–72. https://doi.org/10.1177/1080569908321472

Guba, E. G., & Lincoln, Y. S. (1994). Competing paradigms in qualitative research. In N. K. Denzin & Y. S. Lincoln (Eds.), *Handbook of qualitative research* (pp. 105–117). Sage.

Haas, A. M., & Eble, M. F. (2018). Introduction: The social justice turn. In A. M. Haas & M. F. Eble (Eds.), *Key theoretical frameworks: Teaching technical communication in the twenty-first century* (pp. 3–19). Utah State University Press.

Haas, A. M., & Frost, E. A. (2017). Toward an apparent decolonial feminist rhetoric of risk. In D. G. Ross (Ed.), *Topic-driven environmental rhetoric* (pp. 168–186). Taylor & Francis.

Halloran, S. M. (1982). Aristotle's concept of ethos, or if not his somebody else's. *Rhetoric Review, 1*(1), 58–63. https://doi.org/10.1080/07350198209359037

Hargrave, T. J., & Van de Ven, A. H. (2017). Integrating dialectical and paradox perspectives on managing contradictions in organizations. *Organization Studies, 38*(3–4), 319–339. https://doi.org/10.1177/0170840616640843

Hartelius, E. J., & Browning, L. D. (2008). The application of rhetorical theory in managerial research: A literature review. *Management Communication Quarterly, 22*(1), 13–39. https://doi.org/10.1177/0893318908318513

Heath, S. B. (1983). *Ways with words: Language, life, and work in communities and classrooms.* Cambridge University Press.

Henderson, A., Cheney, G., & Weaver, C. K. (2015). The role of employee identification and organizational identity in strategic communication and organizational issues management about genetic modification. *International Journal of Business Communication, 52*(1), 12–41. https://doi.org/10.1177/2329488414560278

Heracleous, L., & Barrett, M. (2001). Organizational change as discourse: Communicative actions and deep structures in the context of information technology implementation. *Academy of Management Journal, 44*(4), 755–778. https://doi.org/10.5465/3069414

Heracleous, L., & Klaering, L. A. (2014). Charismatic leadership and rhetorical competence: An analysis of Steve Jobs's rhetoric. *Group and Organization Management, 39*(2), 131–161. https://doi.org/10.1177/1059601114525436

Herder, R. (2015). Well-timed stories: Rhetorical *Kairos* and antenarrative theory. In D. M. Boje (Ed.), *Storytelling and the future of organizations: An antenarrative handbook* (pp. 347–365). Taylor & Francis.

Herndl, C. G., Hopton, S. B., Cutlip, L., Polush, E. Y., Cruse, R., & Shelley, M. (2018). What's a farm? The languages of space and place. In C. Rai & C. G. Druschke (Eds.), *Field rhetoric: Ethnography, ecology, and engagement in the places of persuasion* (pp. 61–94). The University of Alabama Press.

Herndl, C. G., & Licona, A. C. (2007). Shifting agency: Agency, *Kairos*, and the possibilities of social action. In M. Zachry & C. Thralls (Eds.), *Communicative practices in workplaces and the professions: Cultural perspectives on the regulation of discourse and organizations* (pp. 133–153). Baywood Publishing Company.

Higgins, C., & Walker, R. (2012). Ethos, logos, pathos: Strategies of persuasion in social/environmental reports. *Accounting Forum, 36*(3), 194–208. https://doi.org/10.1016/j.accfor.2012.02.003

The Holy Bible, New International Version. (2002). Zondervan. (Original work published 1973).

Hyland, K. (2012). *Disciplinary identities: Individuality and community in academic discourse.* Cambridge University Press.

Isaksson, M., & Jørgensen, P. E. F. (2010). Communicating corporate ethos on the web: The self-presentation of PR agencies. *Journal of Business Communication, 47*(2), 119–140. https://doi.org/10.1177/0021943610364516

Jaworska, S. (2018). Change but no climate change: Discourses of climate change in corporate social responsibility reporting in the oil industry. *International Journal of Business Communication, 55*(2), 194–219. https://doi.org/10.1177/2329488417753951

Jones, N. N. (2016). Narrative inquiry in human-centered design: Examining silence and voice to promote social justice in design scenarios. *Journal of Technical Writing and Communication, 46*(4), 471–492. https://doi.org/10.1177/0047281616653489

Jones, N. N., Moore, K. R., & Walton, R. (2016). Disrupting the past to disrupt the future: An antenarrative of technical communication. *Technical Communication Quarterly, 25*(4), 211–229. https://doi.org/10.1080/10572252.2016.1224655

Jones, N. N., & Walton, R. (2018). Using narratives to foster critical thinking about diversity and social justice. In A. M. Haas & M. F. Eble (Eds.), *Key theoretical frameworks: Teaching technical communication in the twenty-first century* (pp. 241–267). Utah State University Press.

Jørgensen, K. M. (2015). Antenarrative writing: Tracing and representing living stories. In D. M. Boje (Ed.), *Storytelling and the future of organizations: An antenarrative handbook* (pp. 284–297). Taylor & Francis.

Killingsworth, M. J., & Palmer, J. S. (2012). *Ecospeak: Rhetoric and environmental politics in America.* Southern Illinois University Press.

Kohn, L. (2015). How professional writing pedagogy and university-workplace partnerships can shape the mentoring of workplace writing. *Journal of Technical Writing and Communication, 45*(2), 166–188. https://doi.org/10.1177/0047281615569484

Kuhn, T. (2006). A "demented work ethic" and a "lifestyle firm": Discourse, identity, and workplace time commitments. *Organization Studies, 27*(9), 1339–1358. https://doi.org/10.1177/0170840606067249

Kvale, S. (1996). *Interviews: An introduction to qualitative research interviewing.* Sage.

Kwortnik, R. J., & Ross, W. T. (2007). The role of positive emotions in experiential decisions. *International Journal of Research in Marketing, 24*(4), 324–335. https://doi.org/10.1016/j.ijresmar.2007.09.002

Latour, B. (1993). *We have never been modern.* Harvard University Press.

LeFevre, K. B. (1987). *Invention as a social act.* Southern Illinois University Press.

Lehtimäki, H., Kujala, J., & Heikkinen, A. (2011). Corporate responsibility in communication: Empirical analysis of press releases in a conflict. *Business Communication Quarterly, 74*(4), 432–449. https://doi.org/10.1177/1080569911424203

Le Rouge, M. (2022). Public responses to a proposed wind farm and their application to technical communication methods. In S. Stinson & M. Le Rouge (Eds.), *Embodied environmental risk in technical communication* (pp. 148–168). Taylor & Francis.

Le Rouge, M., & Stinson, S. (Eds.). (2022). *Embodied environmental risk in technical communication.* Taylor & Francis.

Lester, L. (2010). *Media and environment: Conflict, politics and the news.* Polity.

Lindeman, N. (2013). Subjectivized knowledge and grassroots advocacy: An analysis of an environmental controversy in northern California. *Journal of Business and Technical Communication, 27*(1), 62–90. https://doi.org/10.1177/1050651912448871

MacGregor, S. (2021). Gender matters in environmental justice. In B. Coolsaet (Ed.), *Environmental justice: Key issues* (pp. 234–248). Routledge.

Mackiewicz, J. (2010). The co-construction of credibility in online product reviews. *Technical Communication Quarterly, 19*(4), 403–426. https://doi.org/10.1080/10572252.2010.502091

MacNealy, M. S. (1999). *Strategies for empirical research in writing.* Allyn and Bacon.

Mangum, R. T. (2021). Amplifying indigenous voices through a community of stories approach. *Technical Communication, 68*(4), 56–73.

Martinez, A. Y. (2020). *Counterstory: The rhetoric and writing of critical race theory.* National Council of Teachers of English.

Martinez, D. (2022). Evaluating ecological perceptions and approaches in the fourth national climate assessment report. In S. Stinson & M. Le Rouge (Eds.), *Embodied environmental risk in technical communication* (pp. 169–186). Taylor & Francis.

Martínez, J. G. (2012). Recognition and emotions: A critical approach on education. *Procedia—Social and Behavioral Sciences, 46*, 3925–3930. https://doi.org/10.1016/j.sbpro.2012.06.173

McCormack, K. C. (2014). Ethos, pathos, and logos: The benefits of Aristotelian rhetoric in the courtroom. *Washington University Jurisprudence Review, 7*(1), 131–155. https://openscholarship.wustl.edu/law_jurisprudence/vol7/iss1/9

Meisenbach, R. J., & Feldner, S. B. (2011). Adopting an attitude of wisdom in organizational rhetorical theory and practice: Contemplating the ideal and the real. *Management Communication Quarterly, 25*(3), 560–568. https://doi.org/10.1177/0893318911409548

Mooney, S., Lavallee, S., O'Dwyer, J., Majury, A., & Hynds, P. (2022). Private groundwater contamination and integrated risk communication. In S. Stinson & M. Le Rouge (Eds.), *Embodied environmental risk in technical communication* (pp. 119–147). Taylor & Francis.

Moore, K. R., Jones, N. N., & Walton, R. (2021). Contextualizing the 4Rs heuristic with participant stories. *Technical Communication, 68*(4), 8–25.

Neely, A. D., & Marone, V. (2016). Learning in parking lots: Affinity spaces as a framework for understanding knowledge construction in informal settings. *Learning, Culture and Social Interaction, 11*, 58–65. https://doi.org/10.1016/j.lcsi.2016.05.002

Nowotny, H. (2003). Democratising expertise and socially robust knowledge. *Science and Public Policy, 30*, 151–156. https://doi.org/10.3152/147154303781780461

Olman, L., & DeVasto, D. (2020). Hybrid collectivity: Hacking environmental risk visualization for the Anthropocene. *Communication Design Quarterly, 8*(4), 15–28. https://doi.org/10.1145/3431932.3431934

Paltridge, B. (2012). *Discourse analysis: An introduction* (2nd ed.). Bloomsbury.

Paré, A. (2014). Rhetorical genre theory and academic literacy. *Journal of Academic Language & Learning, 8*(1), 83–94.

Pasztor, S. K. (2019). Exploring the framing of diversity rhetoric in "top-rated in diversity" organizations. *International Journal of Business Communication, 56*(4), 455–475. https://doi.org/10.1177/2329488416664175

Patterson, R., & Lee, R. (1997). The environmental rhetoric of "balance": A case study of regulatory discourse and the colonization of the public. *Technical Communication Quarterly, 6*(1), 25–40. https://doi.org/10.1207/s15427625tcq0601_3

Petersen, E. J. (2018). Female practitioners' advocacy and activism: Using technical communication for social justice goals. In G. Y. Agboka & N. Matveeva (Eds.), *Citizenship and advocacy in technical communication: Scholarly and pedagogical perspectives* (pp. 3–22). Taylor & Francis.

Peterson, T. R. (1997). *Sharing the Earth: The rhetoric of sustainable development*. University of South Carolina Press.

Pickering, K. (2018). Learning the emotion rules of communicating within a law office: An intern constructs a professional identity through emotion management. *Business and Professional Communication Quarterly, 81*(2), 199–221. https://doi.org/10.1177/2329490618756902

Pilger, J. (1998). *Hidden agendas*. Vintage.

Pollach, I. (2018). Issue cycles in corporate sustainability reporting: A longitudinal study. *Environmental Communication, 12*(2), 247–260. https://doi.org/10.1080/17524032.2016.1205645

Prelli, L. J., & Winters, T. S. (2009). Rhetorical features of green evangelicalism. *Environmental Communication, 3*(2), 224–243. https://doi.org/10.1080/17524030902928785

Rai, C., & Druschke, C. G. (2018). On being there: An introduction to studying rhetoric in the field. In C. Rai & C. G. Druschke (Eds.), *Field rhetoric: Ethnography, ecology, and engagement in the places of persuasion* (pp. 1–21). The University of Alabama Press.

Rea, E. A. (2021). "Changing the face of technology": Storytelling as intersectional feminist practice in coding organizations. *Technical Communication, 68*(4), 26–39.

Reynolds, N. (1993). Ethos as location: New sites for understanding discursive authority. *Rhetoric Review, 11*(2), 325–338. https://www.jstor.org/stable/465805

Ross, D. G. (2013). Common topics and commonplaces of environmental rhetoric. *Written Communication, 30*(1), 91–131. https://doi.org/10.1177/0741088312465376

Ross, D. G. (2017). Introduction. In D. G. Ross (Ed.), *Topic-driven environmental rhetoric* (pp. 1–21). Taylor & Francis.

Sackey, D. J. (2018). An environmental justice paradigm for technical communication. In A. M. Haas & M. F. Eble (Eds.), *Key theoretical frameworks: Teaching technical communication in the twenty-first century* (pp. 138–160). Utah State University Press.

Saldaña, J. (2016). *The coding manual for qualitative researchers* (3rd ed.). Sage.

Salvo, M. J. (2006). Rhetoric as productive technology: Cultural studies in/as technical communication methodology. In J. B. Scott, B. Longo, & K. V. Wills (Eds.), *Critical power tools: Technical communication and cultural studies* (pp. 219–240). State University of New York Press.

Sauer, B. A. (2003). *The rhetoric of risk: Technical documentation in hazardous environments.* Lawrence Erlbaum.

Sauer, B. A. (2010). *The rhetoric of risk.* Routledge.

Sauer, B. A. (2022). Reconciling gestures: Overcoming obstacles to transcultural risk communication in South African coal mines. In S. Stinson & M. Le Rouge (Eds.), *Embodied environmental risk in technical communication* (pp. 189–210). Taylor & Francis.

Schryer, C. F., Lingard, L., & Spafford, M. (2007). Regularized practices: Genres, improvisation, and identity formation in health-care professions. In M. Zachry & C. Thralls (Eds.), *Communicative practices in workplaces and the professions: Cultural perspectives on the regulation of discourse and organizations* (pp. 21–44). Baywood Publishing Company.

Schweizer, S., Davis, S., & Thompson, J. L. (2013). Changing the conversation about climate change: A theoretical framework for place-based climate change engagement. *Environmental Communication, 7*(1), 42–62. https://doi.org/10.1080/17524032.2012.753634

Scott, J. B., Longo, B., & Wills, K. V. (2006). Why cultural studies? Expanding technical communication's critical toolbox. In J. B. Scott, B. Longo, & K. V. Wills (Eds.), *Critical power tools: Technical communication and cultural studies* (pp. 1–19). State University of New York Press.

Senda-Cook, S., Middleton, M. K., & Endres, D. (2018). Rhetorical cartographies: (Counter) mapping urban spaces. In C. Rai & C. G. Druschke (Eds.), *Field rhetoric: Ethnography, ecology, and engagement in the places of persuasion* (pp. 95–119). The University of Alabama Press.

Shamir, B., Zakay, E., Breinin, E., & Popper, M. (1998). Correlates of charismatic leader behavior in military units: Subordinates' attitudes, unit characteristics, and superiors' appraisals of leader performance. *The Academy of Management Journal, 41*(4), pp. 387–409. https://doi.org/10.5465/257080

Shim, K., & Kim, J.-N. (2021). The impacts of ethical philosophy on corporate hypocrisy perception and communication intentions toward CSR. *International Journal of Business Communication, 58*(3), 386–409. https://doi.org/10.1177/2329488417747597

Simmons, W. M. (2007). *Participation and power: Civic discourse in environmental policy decisions.* State University of New York Press.

Small, N. (2017). (Re)kindle: On the value of storytelling to technical communication. *Journal of Technical Writing and Communication, 47*(2), 234–253. https://doi.org/10.1177/0047281617769206

Smith, J. M., & van Ierland, T. (2018). Framing controversy on social media: #NoDAPL and the debate about the Dakota access pipeline on twitter. *IEEE Transactions on Professional Communication, 61*(3), 226–241. https://doi.org/10.1109/TPC.2018.2833753

Smith, W. R., Treem, J., & Love, B. (2020). When failure is the only option: How communicative framing resources organizational innovation. *International Journal of Business Communication*. Advance online publication. https://doi.org/10.1177/2329488420971693

Stephens, S. H., & Richards, D. P. (2020). Story mapping and sea level rise: Listening to global risks at street level. *Communication Design Quarterly, 8*(1), 5–18. https://doi.org/10.1145/3375134.3375135

Strauss, A. L. (1987). *Qualitative analysis for social scientists*. Cambridge University Press.

Suddaby, R., & Greenwood, R. (2005). Rhetorical strategies of legitimacy. *Administrative Science Quarterly, 50*(1), 35–67. https://doi.org/10.2189/asqu.2005.50.1.35

Sveningsson, S., & Alvesson, M. (2003). Managing managerial identities: Organizational fragmentation, discourse and identity struggle. *Human Relations, 56*(10), 1163–1193. https://doi.org/10.1177/00187267035610001

Syed, J., & Boje, D. M. (2015). Antenarratives of negotiated diversity management. In D. M. Boje (Ed.), *Storytelling and the future of organizations: An antenarrative handbook* (pp. 47–66). Taylor & Francis.

Taylor, D. (2014). *The state of diversity in environmental organizations: Mainstream NGOs, foundations, government agencies*. Green 2.0. https://orgs.law.harvard.edu/els/files/2014/02/FullReport_Green2.0_FINALReducedSize.pdf

Thomas, G. F., Zolin, R., & Hartman, J. L. (2009). The central role of communication in developing trust and its effect on employee involvement. *Journal of Business Communication, 46*(3), 287–310. https://doi.org/10.1177/0021943609333522

Tillery, D. (2006). The problem of nuclear waste: Ethos and scientific evidence in a high-stakes public controversy. *IEEE Transactions on Professional Communication, 49*(4), 325–334. https://doi.org/10.1109/TPC.2006.885868

Tillery, D. (2017). Scientist as hero, technology as the enemy: Commonplaces about science in environmental discourses. In D. G. Ross (Ed.), *Topic-driven environmental rhetoric* (pp. 43–61). Taylor & Francis.

Tillery, D. (2019). *Commonplaces of scientific evidence in environmental discourses*. Taylor & Francis.

van Dijk, T. A. (1995). Discourse semantics and ideology. *Discourse & Society, 6*(2), 243–289. https://doi.org/10.1177/0957926595006002006

Verboven, H. (2011). Communicating CSR and business identity in the chemical industry through mission slogans. *Business Communication Quarterly, 74*(4), 415–431. https://doi.org/10.1177/1080569911424485

Waller, R. L., & Conaway, R. N. (2011). Framing and counterframing the issue of corporate social responsibility: The communication strategies of Nikebiz.com. *Journal of Business Communication, 48*(1), 83–106. https://doi.org/10.1177/0021943610389752

Waller, R. L., & Iluzada, C. L. (2020). *Blackfish* and SeaWorld: A case study in the framing of a crisis. *International Journal of Business Communication, 57*(2), 227–243. https://doi.org/10.1177/2329488419884139

Walsh, L. (2010). The common topoi of STEM discourse: An apologia and methodological proposal, with pilot survey. *Written Communication, 27*(1), 120–156. https://doi.org/10.1177/0741088309353501

Walsh, L., & Boyle, C. (2017). From intervention to invention: Introducing topological techniques. In L. Walsh & C. Boyle (Eds.), *Topologies as techniques for a post-critical rhetoric* (pp. 1–16). Palgrave Macmillan.

Walsh, L., & Walker, K. C. (2016). Perspectives on uncertainty for technical communication scholars. *Technical Communication Quarterly, 25*(2), 71–86. https://doi.org/10.1080/10572252.2016.1150517

Walton, R. (2013). How trust and credibility affect technology-based development projects. *Technical Communication Quarterly, 22*(1), 85–102. https://doi.org/10.1080/10572252.2013.726484

Walton, R., Moore, K. R., & Jones, N. N. (2019). *Technical communication after the social justice turn: Building coalitions for action.* Taylor & Francis.

Weaver, R. M. (1970). *Language is sermonic: Richard M. Weaver on the nature of rhetoric.* Louisiana State University Press.

Weick, K. E. (1995). *Sensemaking in organizations.* Sage.

Weresh, M. H. (2012). Morality, trust, and illusion: Ethos as relationship. *Legal Communication and Rhetoric: Journal of the Association of Legal Writing Directors, 9,* 229–272. https://papers.ssrn.com/sol3/papers.cfm?abstract_id=2151593

Williams, C. D., Fuson, J. T., & Baker, R. (2016). *Under the lake.*

Williams, M. F., & James, D. D. (2009). Embracing new policies, technologies, and community partnerships: A case study of the city of Houston's bureau of air quality control. *Technical Communication Quarterly, 18*(1), 82–98. https://doi.org/10.1080/10572250802437515

Wynn, J., & Walsh, L. (2013). Emerging directions in science, publics, and controversy. *Poroi, 9*(1), 1–5.

Yates, J., & Orlikowski, W. (2007). The PowerPoint presentation and its corollaries: How genres shape communicative action in organizations. In M. Zachry & C. Thralls (Eds.), *Communicative practices in workplaces and the professions: Cultural perspectives on the regulation of discourse and organizations* (pp. 67–91). Baywood Publishing Company.

Zachry, M., & Thralls, C. (Eds.). (2007). *Communicative practices in workplaces and the professions: Cultural perspectives on the regulation of discourse and organizations.* Baywood Publishing Company.

INDEX

Affordable Connectivity Program, 46
after-action reviews, 72, 217
Alaska, 96, 152
Allen, M. W., 131
antenarrative, 12, 25, 29–30, 197; definition of, 28–29; as negotiating, 30, 34, 64, 162, 187, 201–202, 210; as resistance to dominant narratives and stories, 64, 78, 93, 110, 112, 148, 175
Aristotle: character, 35–36, 127, 129; co-construction of ethos, 153; credibility, 35; identification, 39; trust, 36, 42
Arnett, R. C., 26
attributes, source relational, 37

baptism, 3, 49–51, 54, 153; as part of community value of religion, 50–51, 153, 189
bass, 4
Baumlin, J. S., 36
Bible Belt, 50
Boje, D. M., 29
Bourdieu, P., 13, 30–31, 64
Braet, A. C., 36
Browning, L. D., 44, 68, 80, 133, 205
Burke, K., 34, 39
Burkean, 39
Butts, S., 46, 144
Byrd, Frank, 158, 160

California, 63, 96, 99–100, 103–104, 107, 109, 115
Campbell, George, 110
Campbell, K. K., 36
Caretta, M. A., 74
Carlson, E. B., 74
catfish, 4
Chamber of Commerce, 118, 158, 171–172
character. *See* ethos: and appeals to character
Cherokee, 28
cleanup/clean-up, 151, 163, 165, 183
closure, 76, 83–84, 94, 102, 112, 129, 135, 163, 169, 195, 217; as a Corps plan, 101, 117–118; goals of, 117, 119; imminent, 138; implications of, 5; inevitability of, 64; as an option, 31, 76, 83, 88, 127, 130, 156, 178, 187, 208, 217; possibility or potential for, 4, 83–84, 87, 104, 178; problems caused by, 117–119; and proposed, 107–108, 118; and protect, 65; recommendation for, 69; resistance to, 6, 62, 74, 76, 175; and support of, 53, 58
co-constructing, 6, 32, 41, 153, 187; agency, 186; communication, 18; counterframe, 175; ethos, 14, 19, 37, 79, 93, 116, 152, 194, 206, 210, 218; narrative, 26, 31, 37, 41, 185, 202; solutions, 8, 194; values, 35, 38, 40. *See also* ethos: co-construction of
community. *See* ethos: and community; Grey Cliffs: and community; identity: and community; narrative: community; stories: community; trust: and community; voices, community
Community Eligible Provision, 46
compliance, 40, 175, 177, 189, 198, 203, 211; agency, 154; alignment, 189–190; out of alignment, 68; and convergence, 190, 193, 208; and Corps/government regulations, 91, 124, 134, 139, 149, 175, 185, 216; and enforcing, 68; evidence, 33; negotiating, 185, 189; noncompliance, 145, 185; and organizational goal, 27; and relationships, 21, 122, 175, 193
Conaway, R. N., 40
confirmation bias, 78
conservation, 152–153
contextual model, 197
Convention of States: literature, 17, 20, 86–87, 89, 92–93, 106–107, 110; mindset, 92; organization, 86, 96, 106, 114; website, 106
convergence, 189, 198. *See also* compliance: and convergence
Corps of Engineers. *See* U.S. Army Corps of Engineers
counterframe. *See* framing
counterstory, 28, 45, 61, 112; and antenarrative, 12, 29, 110, 187
crappie, black, 4
credibility. *See* ethos: and appeals to credibility
critical race theory, 28
Cyphert, D., 218

decision making, 45–46, 65, 115
DeVasto, D., 43, 140, 213
dialogue: and cleanup, 165; collaboration, 22, 140, 149, 205; compromise, 40; with organizations, 200; and values, 87, 189
disrespect, 58, 61
Dollar General, 54
Durá, L., 73

elevator pitch, 96, 152–153; creating, 113, 151; developed or developing, 89, 95–96; and Edwards, 96, 153; shaping, 95; sharing, 95–96, 153
emotion: appeals to, 89, 90, 108; about Grey Cliffs closure, 5; and environmental concerns, 47; negotiated, 81; problems with, 90; rejection of, 20, 87–90, 108, 110; unemotional, 24
environment: and communication, 11, 170; damage, 9; and Edwards, 150; embodied or embodiment of, 38, 43; local, 82, 116, 167; natural, 121, 193; nonhuman, 25, 32, 140, 213; and organizational, 48; and physical, 159; and risk, 197; sustainable, 72, 159; and values, 27, 38, 117. *See also* Grey Cliffs: and environment; off-roading: and environmental damage
environmental justice. *See* justice: environmental
environmentalism, 79
erosion, 149; and off-roading, 3, 75, 166, 168; and water quality, 72, 159
ethos: and appeals to character, 10, 21, 35, 37, 42, 44, 85, 110, 116, 124–131, 148, 174; and appeals to credibility, 10, 14, 35, 56, 69–70, 72, 76, 78, 84–86, 92, 98, 101, 110, 112, 116, 122, 125, 148, 174; co-construction of, 6, 8, 10, 14, 19–21, 25, 36–38, 40, 48, 69, 79, 91, 116, 123, 137, 148, 152–153, 164, 194–195, 202, 206, 210, 218; and community, 34, 44, 49, 64–65, 79–80, 123, 164, 175, 195, 200, 211–212, 216; and Dan, 152, 191–192; definition of, 35–36, 153, 195, 211; and Edwards, 21, 37, 44, 56, 64–65, 69, 72, 75–79, 116, 122–131, 133, 137, 139, 143, 148, 152, 160, 174, 195, 212; environmental, 139; invented, 36–37, 133; and narrative, 6, 10, 14, 21–22, 24–25, 34–35, 37–38, 64, 75, 147, 154, 185, 206; as negotiated and negotiation, 6, 10, 14, 22–23, 34, 39, 41, 44, 49, 79, 81, 84–85, 100, 105, 123–124, 130, 133, 142–143, 147–148, 152, 154, 156, 160, 164–166, 175, 186, 194–195, 202, 211–212, 216, 218; and Norma, 20, 81, 84–86, 92–93, 96, 98, 100–101, 104–105, 110, 112, 114, 116; and Paul, 63, 165; and relationship development, 6, 25, 34–37, 69, 79–80, 125; and situated, 37, 133; as spatial metaphor, 211; and Tom, 152, 192; value alignment, 35, 69, 85, 143, 148, 185, 194, 196, 206, 211–212; value frames, 10, 35, 40, 147–148, 186, 193–194, 206; and value terms, 130; and values, 6, 14, 64, 81, 131, 133, 147, 196, 200, 218. *See also* sustainability: and ethos; trust: and ethos
ethos building, 195; and narratives, 6, 19, 22, 25; as a rhetorical framework and theoretical lens, 6, 16, 19, 22, 24, 34, 44; and stories, 19, 25; and values, 6, 22, 40
ethos development, 10, 164; and Edwards, 12, 19, 125; and narrative, 19; and Norma, 98; and process, 152; and values, 8, 194
ethos negotiation, 6, 14, 22, 152, 154. *See also* ethos: as negotiated and negotiation
Eubanks, P. E., 78, 205
expertise, democratized, 78
Express Scribe, 16–17

Faber, B. D.: and agency, 33; and capital, 31; and conflict, 7; and identity-stories, 27; and narrative, 200; and organizational change, 210; and routinization, 209; and self-reflection, 215–216; and stories, 148
Facebook, 86, 97–98, 103, 111, 165, 169–170, 172, 174, 184, 203
Fairhurst, G. T., 210
Fernheimer, J. W., 39
frame. *See* framing
framing: as co-constructed, 8, 41, 175, 186–187, 193; and common interests, 194; and communication 7; as counterframe and counterframing, 31, 175, 185–187, 206; and environmental justice, 22, 45; and ethos, 206; as intentional, 41; and narrative, 9, 21, 31, 41, 149, 184–187, 209; as negotiated, 186, 193; as opportunity and as a positive future, 21, 40, 175, 182–183, 188, 191–193, 207–209; and organizations, 40; as a process, 40, 122–123, 186–187; and rhetorical situation, 116; and risk, 213; and scientific and environmental communicators, 40; and sensemaking, 8–9,

209; and social justice, 22, 42; and solutions, 115, 149; technical and scientific communicators, 40; and theoretical, 19; and values, 25, 35, 38, 40–41, 42, 123, 166, 186–187, 193, 206, 209. *See also* ethos: value frames

George, B., 32, 37, 39
Gephart, R. P., 15
Giddens, A., 33, 149, 185, 209, 216
Goggin, P. N., 134
government: as in antigovernment sentiment, 12, 58, 80, 85, 106, 189; as organization, 9–10, 65, 127, 131, 176, 182–183, 190; skepticism of, 19, 50, 52, 56, 64, 190
Grabill, J. T., 32
Grey Cliffs: and access, 8, 57, 164, 168, 186, 193, 206; and community, 3–4, 6, 11, 17, 22, 27–29, 40, 44, 46, 48, 50–51, 54–56, 58, 72, 74, 77–78, 80, 99, 101, 104–105, 113, 122, 136, 139, 143, 147, 149, 157, 159, 161–162, 166, 175, 185, 189–190, 193, 209, 214, 217; and conflict, 7–8, 30, 213; and crime, 3–5, 27, 56–57, 67, 71, 75–77, 80, 104, 113, 136, 155, 159, 192, 208; and Dan, 95, 182; and Denise, 50, 56, 63, 99, 113, 168; and environment, 3–4, 26, 36, 44–45, 48, 70, 77, 80, 103, 116, 133, 136, 139, 150, 157, 159, 166, 208, 212–213; and Edwards, 3, 5–6, 17, 19, 24, 27, 33, 35, 58, 65, 67, 70–76, 78, 83, 87–88, 94–95, 104, 122, 126–128, 130–131, 133–134, 136–138, 142–143, 147–150, 154–156, 158–160, 165, 170, 174–185, 187, 189, 192, 194, 206, 208, 212, 216–218; and Felicia, 56; as lake-access point, 4–5, 19, 67, 77, 168, 192; and Lee, 53, 57, 63, 167–168; as nefarious, 52, 67, 71, 77; and Norma, 5, 84, 86–88, 90, 92–94, 97–105, 107, 109, 111–113, 116–122, 174; as notorious, 5, 68, 135; and Paul, 54, 171, 173; as positive, 8, 21, 129–130, 134, 159–160, 162, 180, 191–193, 208–209, 212; and preserving, 19, 35, 54, 73, 134, 151, 164, 171, 184, 193; as space, 3, 28, 35, 64, 140, 211; and Tom, 18, 50, 53, 56, 95, 102, 113–114, 136, 150, 152, 155, 157–158, 160, 168–170, 182, 203; and Trisha, 167. *See also* closure
Griffin, F., 42

Halloran, S. M., 36
Hartelius, E. J., 44, 68, 80, 133, 205
Heath, S. B., 61, 99–100

Herndl, C. G., 24, 32–33, 124, 185, 200
Higgins, C., 12
historicity, 11, 26

identity, 214; and agency, 16, 34, 200; co-constructed or shared, 8, 14, 37, 39, 43, 80, 128, 133–134, 137; and community, 25, 28, 79, 133, 205; definition of, 25; as image, 12, 200; and narratives, 25–26; as negotiation, 9, 11, 14, 16, 43, 80, 133, 142; and organization and the Corps, 12, 25, 75, 133, 175, 205, 211; and stories, 27, 75; and work, 40. *See also* narrative: and identity; rhetor: and identity; stories: identity-stories
Iluzada, C. L., 187
injustice. *See* justice
interactionality, 30
intersectionality, 45–46
intraplay, 29
Isaksson, M., 42

James, D. D., 168, 181
Jones, M., 46, 144
Jones, N. N., 26, 29, 45
Jørgensen, P. E. F., 42
justice, 42; environmental, 18–19, 22, 25–26, 44–47, 139, 150; injustice, 41, 46, 116; procedural, 143; social, 22, 25–26, 34, 42, 44–45, 47, 115, 150, 194

kairos, 195
Killingsworth, M. J., 11

Le Rouge, M., 39, 44
Lee, R., 38
Lester, L., 205
lexis, 130
Licona, A. C., 32–33, 124, 185, 200
Lindeman, N., 46, 78
logos, 153; and Norma, 20, 89, 110, 114

MacGregor, S., 116
Mackiewicz, J., 77–78
Manzo, H., 32, 37, 39
map, 144; deep mapping, 139, 144; digital, 74, 139–140, 144; and Edwards, 17, 21, 73–74, 76, 96, 123, 137–140, 153, 191; and environmental justice, 139; of the lake area, 16, 21, 73–74, 76, 96, 137–139, 153, 191; revised, 17, 21, 123, 138–141; rhetorical context of, 73, 144
Martinez, A. Y., 28
Martinez, D., 32
me attitude, 43
Merck, 42

Meyer, C. A., 36
model, and deficit, 197
Mooney, S., 43
mudslide, 3, 80, 166

narrative: and agency, 6, 28, 32–35, 41, 47, 149–150, 162, 165, 185, 195–196, 200; and capital, 30–31, 34–35, 64, 143, 150; changed, 21; co-constructed or shared, 7, 21, 25–26, 31–33, 37–38, 40–42, 48, 99, 149–150, 154, 157, 162, 166, 175, 177, 179, 182, 185, 187, 201–202; community, 6–7, 9, 11–12, 15, 24–25, 27, 30–31, 48, 65, 77, 79, 122, 148, 150, 162, 185, 202, 210; and culture, 8, 14, 25, 28, 30–33, 61, 77, 80, 99, 139, 209; definition of, 26; dominant, 9, 25, 28–29, 47, 61, 64, 75–76, 81–82, 148, 195–196, 201; Edwards's, 6, 21, 30–31, 70–71, 75–79, 116, 124, 126, 128, 142–143, 148, 179, 185, 202, 210–211; environmental preservation, 48; as in heteroglossic, 12; and identity, 16, 25–26, 48, 75, 175; negotiated, 7, 16, 30, 80, 147, 165, 185, 206, 213; Norma's, 92; organizational, 25, 29, 34, 48, 75, 84, 141–142, 211; and polarizing, 5–7, 9, 24, 26, 31, 42, 64, 76–79, 81, 123, 130, 148, 186–187; and rhetoric, 15–16, 33, 41, 48, 70, 76, 86, 116, 126, 202, 211; self-narrative, 12, 15–16, 21, 49, 58, 61, 69, 93, 125, 215. *See also* antenarrative; ethos: and narrative; recreation: narratives and stories of
National Environmental Policy Act, 124
newcomers, 99–100, 218
noncompliance, 145, 216
nondominance, 46
nonexpert, 43, 190, 213
nonhuman, 32, 46, 140
nonprofit, 96, 103

off-roading, 129, 150, 177, 180, 194; and environmental damage, 27, 72, 75, 166, 168, 190; off-road vehicle, 67, 79, 118–119, 129, 166, 195; and stories, 27, 75, 186; and trespassing, 53, 167
oldtimers, 99–100
Olman, L., 43, 140, 213
organizing, 164, 179, 208, 216; and Dan, 96; as in grassroots, 5; and Norma, 5, 87–88, 94, 110–112, 120; and texts, 87, 110

Palmer, J. S., 11
Patterson, R., 38
Petersen, E. J., 115

policy, 11, 22, 24, 46–47, 65–66, 81, 117, 144–145
polyphony, 12; in organizational communication 39; of voices, 30, 38–39, 76, 165
power, hydroelectric, 4, 9
Prelli, L. J., 212
preservation, 150, 165, 196, 198, 219
public, 10, 12, 17, 22, 38–39, 41–42, 45, 48, 68, 70, 78, 83, 104, 126, 128, 131–132, 134, 143–144, 149, 157–159, 166, 168–170, 180–182, 184, 187, 190–192, 196–197, 200, 212, 214, 218; and government organizations, 10, 37; and lands, 39, 67–68, 70, 126–127, 131–132, 134–135, 187, 190

Quintilian, 135

recreation, 126, 175, 182; narratives and stories of, 7, 55; value of, 19, 50–51, 55, 64, 153, 187–188, 190, 193
regulation, 125, 137, 218; and compliance, 149; and the Corps, 11, 20–21, 67, 70, 73–74, 91, 103, 122–123, 130, 137–140, 142, 145, 169–170, 173, 176, 178, 185, 202, 206, 209, 216; federal, 108, 145–146; Federal Code of, 67, 70, 75, 145; and government, 134, 196; as in laws and rules, 7, 20–21, 24–25, 27, 46, 61, 64–65, 67, 74, 81–82, 96–97, 103, 105–106, 118–119, 122, 126, 132, 136–141, 145, 169–170, 175, 181, 185–186, 189–191, 196, 202, 205–207, 209, 212, 216; and Norma, 109; of policy, 45, 82. *See also* rhetoric: and regulation
rejuvenation: allowing for, 138; and closing, 77; environmental, 157, 165; plan for, 17, 142; and process, 162
relationship, 7, 19, 32, 57, 61–62, 64, 116–117, 125, 132, 134, 140, 143, 148, 151, 157, 165, 176, 190–191, 193, 205, 207, 209, 217, 219; and agency, 32, 34, 196, 201; and co-constructing and negotiating, 34, 37, 42, 69, 79–80, 111, 138, 212, 218; cultural-historical, 124; and Edwards, 123–125, 128, 131, 136–138, 142, 175–176, 178–179, 185, 188–190, 192; geographic, 73; and Lee, 204; and Norma, 97; and power, 29; problematic, 55; rhetoric of, 19, 25–26, 31, 34–35, 39, 44, 48, 124; and social action, 7, 10, 16, 23, 30, 91, 142; and Tom, 170; and trust, 43, 48, 56, 78, 112, 127, 187, 191, 211. *See also* ethos: and relationship development
religion, and values, 19, 50–51, 64, 187–188, 190, 193

Index   235

resource manager, 3, 9, 12, 16–17, 19, 65, 67–72, 83, 88, 122, 124, 127, 134, 152, 178–179, 184, 193, 208, 217; and authority, 69–70, 72, 124, 143, 160; and power, 33; and U.S. Army Corps of Engineers, 19, 67, 69, 83
respect, 39; and communication, 37; and community, 33, 37, 59–61, 63; and lack of, 58; and need for, 19; and social unity, 50, 52, 58, 60–61, 63; as a value, 57–58, 60–61, 63
Reynolds, N., 153
rhetor, 14, 33, 78, 153, 202; authority, 201; and character, 125; co-construction or negotiated, 36–37, 39, 41, 44; and community, 9, 11; credible, 134; and Edwards, 19, 187; and identity, 39, 80, 205; organizational, 9, 11, 39, 44, 68; regulatory, 137; and trust, 41
rhetoric: and analysis, 15, 35, 84; and appeals, 76, 115; Aristotelian, 35; and competing, 24; cultural, 33, 185; extrarhetorical, 202; and rhetorical and theoretical framework, 21, 24, 48, 69–70, 86, 126–127, 211; and identification and identity, 37, 39, 48, 73; and organizational leaders, 10; and regulation, 69, 73, 124; relationship-building, 134; and *Rhetoric*, 35, 42, 129; and rhetorical situation, 14, 37, 69, 116, 177, 202, 217, 219; strategies, 14, 35, 116, 123, 147; and topoi, 41–42; and values, 8, 10. *See also* ethos: and appeals to character; ethos: and appeals to credibility; ethos building: as a rhetorical framework; framing: and rhetorical situation; map: rhetorical context of; narrative: and rhetoric; relationship: and rhetoric of
risk visualization, 43, 140
Ross, D. G., 41
routinization, 209
runoff, 72, 80, 159

Sackey, D. J., 27, 45, 48
Salvo, M. J., 202, 218
self-reflection, 215, 217
sensemaking, 15, 29, 91, 161, 197, 209–210
sexism, environmental, 115
Simmons, W. M., 47, 197
Small, N., 28
Smith, W. R., 7, 40, 207–211
social construction, 39; social constructionist model, 213
social justice. *See* justice: social
Southern Baptist, 50

space: affinity, 177; as beloved and loved, 3, 205; as contested, 152; geographic, 3, 5, 28, 35, 46, 48, 64, 140, 144, 157, 177, 185, 196, 198, 205, 210–211, 219
Stinson, S., 39
stories: and agency, 30, 32–34, 47, 150, 162, 202; and capital, 30–31, 34, 64; community, 4, 14–15, 19, 21, 24, 27–31, 34, 47–49, 52, 55–57, 61, 64–65, 77, 82, 90, 93, 99, 122–123, 130, 141–142, 144, 148–151, 154, 156–157, 160–162, 184–187, 189, 210, 213–214, 218; and conflict or divergent, 7; and the Corps, 4, 30, 47, 77, 123, 137, 142, 148–149, 151, 160–161, 184, 210; cultural-historical, 27, 31, 99, 218; definition of, 25–27, 29–30; and dominant, 28, 82; and Edwards, 15, 27, 31, 73, 75, 77, 123, 137, 154, 159–160, 162, 184–185, 210, 218; and family time, 4; and framing, 209; identity-stories, 27, 75; living, 29; and Norma, 115; and relationship, 28, 30, 34, 48–49, 218; relationship-building, 28; and social action or change, 30, 32, 35, 148, 210; and Tom, 152, 186; and unifying, 7, 28, 210; and values, 19, 27, 32, 35, 47–49, 57, 61, 64–65, 77, 82, 90, 122–123, 153, 187, 209, 218. *See also* antenarrative; counterstory; identity: and stories; off-roading: and stories
strong-arming, 124
structuration theory, 185, 209
sustainability: and communication, 14; definition of, 134; and Edwards, 19, 128, 134, 188–190; efforts, 133, 170; and environmental justice and responsibility, 45, 197; and ethos, 8, 44; and goals, 9, 11, 133, 187; and issues, 219; and mission, 181; and narrative, 8, 44, 213; and needs, 219; and process, 142; and regulations, 67; and rhetoric, 10; and values, 6, 8, 10, 35, 38, 41, 131, 133–135, 137, 187–188, 189–191, 193, 211

Thomas, G. F., 56
Tillery, D., 41, 46, 79, 125
topoi. *See* rhetoric: and topoi
tradition, 4, 214; and community, 6, 53; and oral, 61; and values, 19, 50–51, 53–54, 64, 187–188, 190, 193–194
trout, 4
trust: and building, 21, 43, 50, 85, 166, 180, 187; capital of distrust, 143; capital of trust, 143; and co-construction and negotiation process, 19, 25, 35, 38, 42–44, 50, 55, 81, 85, 165; and com-

munity, 31, 43–44, 55, 85, 100, 103–104, 112, 123, 127, 143, 154, 166, 187–188, 191; and culture, 39, 41; definition of, 56; development of, 10; distrust, 9, 56–57, 80–81, 97, 104, 112, 113, 156, 189; and Edwards, 31, 44, 80, 123, 127–128, 135, 143, 154–155, 166, 180, 187–188, 191; entrustment, 135, 188; and ethos, 19, 42, 44, 85, 112; and identity, 39, 43; and Norma, 81, 100, 103, 112, 116; and Tom, 155; untrustworthiness, 85; and values, 39, 41–42, 48, 50, 55, 85, 114, 123, 134–135, 137, 143, 154, 165–166, 188, 191, 211. *See also* Aristotle: trust; relationship: and trust

U.S. Army Corps of Engineers, 24, 110, 145–146; the Corps, 16, 21–22, 26, 32, 46, 48, 57, 68, 72, 79, 100, 129, 132, 142, 148–149, 154, 158–161, 165, 176, 184, 189, 192–193, 201, 203, 207, 209, 213. *See also* closure: as a Corps plan; compliance: and Corps/government regulations; identity: and organization and the Corps; regulation: and the Corps; resource manager: and U.S. Army Corps of Engineers; stories: and the Corps

unity, social, 19, 50, 52, 58, 61, 64, 188. *See also* respect: and social unity

values, and terms, 126–128, 130–132, 133–134, 212. *See also* co-constructing: values; dialogue: and values; environment: and values; ethos: and values; ethos: value alignment; ethos: value frames; ethos: and value terms; ethos building: and values; ethos development: and values; framing: and values; recreation: value of; religion: and values; respect: as a value; rhetoric: and values; stories: and values; sustainability: and values; tradition: and values; trust: and values

Vioxx, 42

voices, community, 11, 58, 160, 165

Walker, K. C., 208
Walker, R., 12
Waller, R. L., 40, 186
walleye, 4
Walsh, L., 41, 208
Walton, R., 26, 39, 43–45
Weresh, M. H., 42
Williams, M. F., 168, 181
Winters, T. S., 212

XYZ Company, 163, 181, 185, 215; and collaboration, 136, 158; and Dan, 95, 114, 151, 182, 191; and Edwards, 114, 136, 158, 162, 183, 191–192; and elevator pitch, 89, 95, 113–114; and ethos, 154; and financial support or funds, 95, 154, 159, 185; and ideas, 158; and materials, 95, 154, 168; and plan and purpose, 96; and stories, 162; and texts, 153; and Tom, 95–96, 114, 151, 153, 155, 158, 182, 191

you attitude, 43

ABOUT THE AUTHOR

Kristin Pickering is professor of English at Tennessee Technological University, where she directs the Professional and Technical Communication Program. She earned M.A. and Ph.D. degrees in English—rhetoric and composition from the University of South Carolina in Columbia, as well as a B.A. in English from Columbia College. Her interests in engineering and technical communication stem from graduate work in the Electrical and Computer Engineering Writing Center at the University of South Carolina; she completed her dissertation on written genre knowledge within electrical and computer engineering. At Tennessee Technological University, Dr. Pickering teaches courses in oral communication, ethics, technical editing, technical/professional writing and communication, and research methods at the undergraduate and graduate levels. She has published a number of scholarly articles and book chapters.